Digital Modes for All Occasions

Murray Greenman, ZL1BPU

Radio Society of Great Britain

Published by the Radio Society of Great Britain, 3 Abbey Court, Priory Business Park, Bedford, MK44 3WH. UK

First published 2002

Reprinted 2004, 2006 & 2008

Digitally repinted 2011 onwards

© Radio Society of Great Britain 2002. All rights reserved. No part of this publication ma be reproduced, stored in a retrieval system, or transmitted in any form, or by any means, electronic, mechanical, photocopying, recording or otherwise, without the prior written agreement of the Radio Society of Great Britain.

ISBN 9781 8723 0982 8

Publisher's note
The opinions expressed in this book are those of the author and not necessarily those of the RSGB. While the information presented is believed to be correct, the author, publisher and their agents cannot accept responsibility for consequences arising from any inaccuracies or omissions.

Cover design: Anne McVicar
Illustrations: Murray Greenman
Typography: Mike Dennison, G3XDV, of Emdee Publishing
Production: Mark Allgar, M1MPA

Printed in Great Britain by Page Bros Ltd of Norwich

Contents

1 Introduction **5**
What is a digital mode? Fuzzy modes. Advantages of digital modes. Disadvantages of digital modes. Introduction to the modes.

2 Digital fundamentals **15**
The digital transmitter. The digital receiver. Considering the data. Data codes.

3 Serial data transmission **25**
Data synchronism. Error correction techniques. Encryption techniques.

4 Digital modes and the ionosphere **31**
Noise and interference. Multipath reception. Doppler modulation effects. Comparative performance. Designing modes for HF performance.

5 History **45**
Text transmission. Automatic reception. Image transmission. Amateurs involved. Amateur designs. Error coding systems.

6 Getting started **53**
What you need. Connecting up. Setting up. Operating hints. Choosing software. Special operating considerations.

7 AmTOR **71**
Error correction. AmTOR mode A. AmTOR mode B. Operating AmTOR. Performance.

8 Clover **77**
Clover II modulation. Error correction. Performance. Clover 2000. Operating Clover.

9	**Hellschreiber** 81
	How Hell works. Feld-Hell. Multi-tone Hell. PSK Hell. Other Hell modes.

10	**MFSK modes** 91
	How MFSK works. MFSK16. MFSK8. Throb. FSK441.

11	**MT63** 103
	Trasnmiting MT63. Receiving MT63. MT63 software.

12	**PacTOR** 109
	PacTOR ARQ. PacTOR FEC. PacTOR II.

13	**PSK31** 115
	How PSK31 works Operating PSK31. Software for PSK31.

14	**RTTY** 121
	History. RTTY described. Using RTTY. Hardware for RTTY. Software for RTTY.

15	**Image modes** 129
	SSTV Fascimile. Satellite images. New image modes.

16	**Advanced digital modes** 137
	PC-ALE. STANAG 4285, 45239. Q15X25. Digital voice communications. Ionospheric soundings.

17	**Other tools for amateurs** 143
	Spectrograms and spectrum analysers. Dopplergrams. Oscilloscopes. Wave tools. Other analysis tools.

APPENDICES

A	Digital modulation techniques 149
B	Alphabet and code reference 161
C	Software and information sources 169
D	The PC sound card 175
E	How the pictures were made 191
F	Glossary 199
G	Bibliography 203
	Index 207

Introduction

In this chapter

- What is a digital mode
- Fuzzy modes
- Advantages of digital modes
- Disadvantages of digital modes
- Introduction to the modes

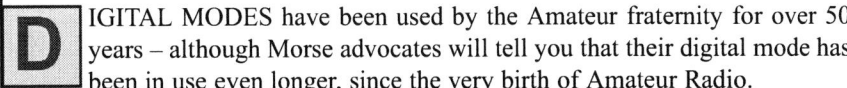

DIGITAL MODES have been used by the Amateur fraternity for over 50 years – although Morse advocates will tell you that their digital mode has been in use even longer, since the very birth of Amateur Radio.

Strictly speaking, a digital mode is one in which the transmission of information takes place using discrete (ie digital) properties of a radio transmission, rather than a smoothly varying (analogue) process. However, this definition is both too strict on the one hand, and too loose on the other, so let's go with a simpler and more inclusive definition – if a computer or terminal is used to transmit or receive the mode, we'll call it a digital mode.

Most Amateur activity has, until recently, used Morse or voice, with other modes forming only a very small proportion of the total activity. Times are now changing, and many digital modes are becoming mainstream activities. Amateur packet radio on VHF has been widely used for 15 years or so; digital modes are used to control and communicate with Amateur satellites; and HF bulletin boards using RTTY, AmTOR and later PacTOR have existed for many years. Now, as packet activity declines due to the wide acceptance of Internet e-mail, HF digital mode operation is stronger than ever and growing daily (**Fig 1.1**). Amateur Radio need not compete with the Internet – indeed the Internet is enhancing the hobby for many digital mode operators. While there is interest in message handling, offering mobile and portable stations access to e-mail (through bulletin boards), the area of strongest growth involves live conversations around the world using new high performance chat modes such as PSK31 and MFSK16.

It is of interest to ponder why, in the past, digital modes have been something of a fringe activity. Maybe it was cost, maybe the complexity of the equipment, maybe an unwillingness to learn something new or complex, and maybe the performance experienced was less than desired. Maybe it was all these things. The

French operators sending and receiving messages on Baudot multiplex telegraphs.

Fig 1.1: 'Alors mes amis – we are in trouble! I read here that someone has invented the Internet!'

scene is now changing very quickly, as the ease of use, versatility and performance of low cost digital systems are improving at an impressive rate. New Digital modes and new software are appearing all the time. Indeed, it has been claimed by physicist and radio-propagation expert Dr Gary Bold ZL1AN that more new radio modes have been developed in the past five years than were developed in the previous 50! [1]

This book is largely about modern digital modes, and deals especially with those that use the PC sound card. However, for completeness, traditional techniques are described as well. In order to understand where the state of the art is heading, and why modern modes are designed in the way they are, it is useful to understand something of the course the technology has taken over the last 50 years to arrive at the state of the art. Some of these traditional modes have recently been brought up to date – Hellschreiber is a spectacular example (**Fig 1.2**) – and justly deserve consideration as modern modes.

To use this technology effectively, it is necessary to understand a little of how each particular mode works, and how to install and operate the software and hardware involved. In order that the reader may achieve the results promised, it is also useful to appreciate the transmitter requirements, and the effects of ionospheric propagation. These things are covered in this book.

However, the reader need not be concerned – there will be no confusing technical detail or high-powered mathematics. Things will be kept simple, and if the reader really wants to know in detail how each mode works, they can refer to the footnotes, the resources mentioned, and the appendices at the back of the book.

Thus the book will not only show you how to install, set up and use the modes, it will give you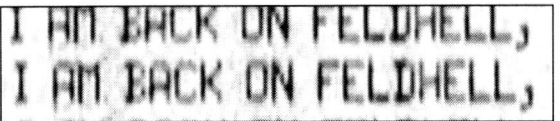

Fig 1.2: Sensitive DX Hellschreiber reception using grey-scale display and DSP techniques

the history of, and background to, digital modes. In describing these modes, the transmitter requirements and HF propagation characteristics that affect the performance are also considered. A good understanding of these characteristics will allow the operator to make the right choice of mode for each occasion. After all, this book is about HF Digital Modes for all Occasions.

What is a Digital Mode?

Many Amateurs are unsure what is meant by 'digital modes', or how to make use of them, and this alone may keep one from enjoying this growing branch of Amateur Radio. There is much more to operating Digital Modes than just packet radio (**Fig 1.3**), or sending messages of doubtful utility or relevance, as the caption to the picture below implies!

Nor are Digital Modes just about RTTY, or bulletin boards. There are real DX contacts going on, real conversations, lots of fun, and new friends being made daily with the various digital modes. The most exciting aspect of using Digital Modes is the 'real time' nature of operating – just like Morse or SSB.

The conventional definition describes Digital Modes as those using discrete properties of a radio transmission to send discrete data, for example turning on and off a carrier to signal a 'one' or a 'zero'. These discrete properties are modified by a modulator of some type, be it an on–off keyer, a frequency shift system, or a phase shifter. The conventional definition is however inadequate and an explanation is in order.

Strictly Digital

The strict definition of a digital mode leads to some confusion, as in all cases a discretely modulated signal is rarely actually transmitted as a purely digital signal, and is in all cases actually received as an analogue signal. How is this?

Fig 1.3. 'Hey John – I got your packet message this morning! Only took three days to arrive!' (Drawing: Bob Knowles)

Well, the transmission is always filtered or shaped in some way, usually to restrict the bandwidth, and always to ensure that harmonics and out-of-band products are not transmitted. So, the signal transmitted is shaped by the filters, and therefore analogue. Next, the ionosphere modulates the transmitted signal, which it can do by changing the time of arrival, by providing multiple signal paths, or by changing the strength and phase of the signal and even the frequency. In receiving a digital signal, further filters are used, resulting in even more shaping of the signal. The process of restoring the received signal to a digital form is an important part of the receiving system, and this decision process must be carefully designed to minimize the number of errors received.

So, hard to believe, but digital transmissions are analogue when transmitted, analogue as they pass through the ionosphere, and analogue as they pass through the receiver!

Loosely Digital

The conventional definition of what makes a digital mode is also too loose, in that there are several transmission types that are recognised as digital (well, they send data from a computer, so they must be digital, right?), but are in fact analogue. For example, SSTV is an analogue mode, as is FAX. Many amateurs enjoy receiving weather satellite pictures (or SSTV pictures from Amateur satellites) with beautifully defined clouds and ground detail. These are undoubtedly analogue transmissions.

For the purposes of this handbook a more inclusive definition will be followed:

A digital mode is one where we use a computer or computer terminal somewhere in the process of transmitting or receiving.

This allows inclusion of SSTV, FAX, and even medium scan TV perhaps, as digital modes.

Now clearly, SSB voice, FM voice and AM are analogue modes. Some modes, like SSTV, have analogue image transmission, yet have digital properties as well, such as transmission of sync.

But what about Morse and Hellschreiber? Morse is clearly a digital transmission, since it is an on and off technique, but the reception is analogue – the skilled operator uses more than just the on–off beeps to interpret the signal. Noise, drift, chirp, click, 'fist', sending speed and keying weight all contribute to the receiving experience.

The same is true of Hellschreiber, where the receiving operator enjoys the hesitations in typing, drift in sending speed, the affects of the ionosphere, and even the different fonts used by the sender.

Fuzzy Modes

This is perhaps a suitable opportunity to introduce another notion – the 'fuzzy' mode. The two modes just discussed, Morse and Hellschreiber, are the best examples of fuzzy modes, in other words modes that are neither fully digital, nor are they analogue. HF FAX (used principally for weather maps) can also be considered a fuzzy mode.

A fuzzy mode is described in terms of three parameters, 'how', 'when' and 'what':

A fuzzy mode is one where:

HOW – **The transmission is essentially digital**
WHEN – **Reception involves no timing decisions in the receiving equipment**
WHAT – **Reception involves no data decisions in the receiving equipment**

While a fuzzy transmission is shaped in an analogue way, as Morse and Hellschreiber are shaped to reduce clicks, it is the digital properties of the transmission (typically amplitude and timing or frequency) which define how the data is transmitted.

The main difference is in the receiver. The fuzzy receiver makes no decisions about when a data element arrives – this is left to the operator. While this may seem obvious, it is an important notion, because it allows the operator to interpret reception by ear or eye, using the great powers of the brain in pattern recognition to correctly place the received data, despite sending idiosyncrasies and ionospheric timing differences. A similar fuzzy process is involved in deciding whether a received Morse element is a dot or dash.

Similarly, the fuzzy receiver makes no decisions about what the data is. Rather than deciding whether a data element is received or not, the receiver presents to the operator the probability that a data element is received. In Morse terms, this is obvious, as a weak dot will still be perceived as a dot by the brain. There are important ramifications here for computer reception of Morse, because the computer has great difficulty interpreting the large variation in dot strengths. In the same way, with Hellschreiber reception, copy is always better if the computer display shows the relative strength of each dot as greyness, without the computer or equipment making any decisions about what is a dot and what is not.

These three properties of a fuzzy system – the how, the when and the what – can be used to assess other modes for 'fuzziness', and to assess computer-based systems for their ability to receive fuzzy modes correctly.

The two main features of fuzzy modes are their disarming simplicity (the digital processing, pattern and syntax recognition are performed in the brain), and their inherent ability to survive levels of interference that would cripple other modes.

Fuzzy modes are highly effective, specifically because there is no rigid synchronism of the data or interpretation of the data.

Advantages of Digital Modes

The advantages of digital modes compared with analogue (typically voice) modes are many, although until recent improvements in technology were embraced, not all of these have been realised in practice. Some modes are better than others in various respects. The main advantages are:

- High sensitivity (not much power required)
- Very robust communications (useful in poor conditions and terrible QRM [2])
- Narrow transmission bandwidth
- Computers can be used, allowing automatic operation, keyboard sending, file transfer, e-mail forwarding and automatic logging
- Error correction techniques can be easily used
- Digital modes are very useful for Amateurs with a hearing disability, speech impediments or poor hand coordination.

Several of these advantages go hand-in-hand. Narrow bandwidth leads to high sensitivity and reduced susceptibility to interference. Several recently-designed digital modes allow users to enjoy really good DX [3] with very modest stations. It is not unusual to have intercontinental or even antipodal [4] QSOs [5] using 5W and a dipole antenna. On the lower bands, contacts over 1000km or more, using 5W or so, have become commonplace. While these performance claims are old news to Morse operators, few voice operators can boast such results.

There is no doubt that the convenience of using a personal computer (PC) for HF operation has great appeal to some. The PC can provide control of the rig (even tuning the rig and changing bands), it can act as the modem, provide digital signal processing, provide a permanent record of the conversation through automatic logging, send and receive files, and act as a gateway from the radio link to the Internet.

In many cases the operator can change modes simply by changing the operating software, and much of the very best software is completely free of charge, easy to use and easily available.

Older digital modes either used no error correction, or required specialized external hardware to allow modest error detection and sometimes correction. Now with the PC involved, it is possible to provide a very high order of error correction without the operator even being aware that it is taking place.

Many of our friends have given up radio, perhaps never started, or admit to not enjoying it anymore, on the grounds that they have trouble operating, hearing or sending, due to some disability or the slow falling apart of their bodies with age. Hearing is not necessary for the use of digital modes. With just the ability to see and type, disabled Amateurs can be active. It is not difficult to learn to use the most popular modes, even at an advanced age.

This book has been written to encourage Amateurs to take up digital modes, enjoy them, and make good use of the many benefits and advantages of these modes.

Disadvantages of Digital Modes

While the use of digital modes has largely positive attributes, there are a few disadvantages that should also be considered. Let's at least pretend to be unbiased! For example:

- Operating portable or mobile isn't easy (you need a modern laptop computer and a source of power for it)
- Digital mode operation can be moderately expensive (you do need a computer, and sometimes have to buy software or hardware)
- The computer can be a serious source of interference on HF
- The ionosphere can effect digital modes more strongly than voice modes
- Operating does require extra skills and knowledge
- Tuning accuracy, drift and transmitter linearity can be very important
- Great care must be taken with the station setup to avoid poor transmission quality, hum, noise, RF feedback and interference from the computer
- Typing skills in excess of 10 WPM [6] are required, and you need to be able to read the computer screen clearly (blind digital mode operators do exist!)

All of these disadvantages can be surmounted, and with the assistance of this book and perhaps a friend to help out, any Amateur with a computer can soon be on the air and operating the new modes with complete success.

Introduction to the Modes

Most of the modes briefly mentioned here have a complete chapter devoted to them later in the book. In each chapter the mode will be described, along with how it is used, what it is best used for, and how to recognise the mode when you meet it.

These are the modes you are most likely to meet on the air. Some modes have multiple names, and in this case the most common name is used. There are of course many more modes than are listed here!

Modes are described as 'chat' modes (used like SSB or Morse), 'connected' modes (used like packet or PacTOR), or 'broadcast', more suited for bulletin transmission. 'Synchronous' connected modes have a precise cycle of transmit and receive.

AmTOR
The name stands for Amateur Teleprinting Over Radio, and the mode was developed in the early 1980s from the commercial SiTOR system, widely used by the marine service. AmTOR was the first Amateur mode to use a dedicated microprocessor system. AmTOR uses FSK [7] modulation at 100 baud [8], and has full-time FEC [9] or ARQ [10] error-correction options, used for chat, broadcast or synchronous connected modes. It is little used now, although the FEC mode is very useful for transmission of bulletins.

Clover
This is a commercial system developed by the HAL Communications Corporation for Amateur and Commercial service. There are several variations, but the most widely used version is Clover II, which uses four sequential tones at 125 baud, phase-shift keyed. Sophisticated error correction and QSO management is offered. The data rate changes to suit the conditions, and it operates in a bi-directional synchronous connected mode. Clover is one of the more expensive systems, but can be very effective under poor conditions.

Facsimile
Commonly known as FAX or HF-FAX, this mode is still widely used commercially for the transmission of weather maps. It is an FSK picture-transmission system, and can take 10–20 minutes to send a single high resolution picture. It has been used to send news, message forms, photographs and even circuit diagrams, but is now little used by Amateurs. The satellite variant, known as SAT-FAX or WEFAX, is popular among enthusiasts, and transmits high-resolution satellite images using a 2400Hz AM subcarrier on an FM transmitter. There is excellent software for Amateur reception of these images, direct from the satellites.

Hellschreiber
Older even than RTTY, Hellschreiber is the second-oldest digital mode we have (after Morse). It was developed in the 1920s and was widely used for press transmissions before, during and after World War Two. After the war it fell into disuse and was largely forgotten, until revived by Dutch and German Amateurs in the 1980s. The original mode (now called Feld-Hell) uses ASK [11] at 122.5 baud, and transmits at a stately 20 WPM. This is still the Hellschreiber mode most widely used, although modern fuzzy software gives much improved reception. There are now a number of high performance DSP [12] variants of Hellschreiber, and the DX performance is up with the best. Most activity is on 80m and 20m.

MFSK16
This unusual chat and broadcast mode owes much to the highly successful Piccolo mode developed in the 1950s by the Diplomatic Wireless Service. MFSK16 uses 16-tone MFSK [13] at 15.625 baud, but with modern techniques such as synchronous FFT [14] detection, phase continuous transmission, and strong full-time error correction, the performance is state-of-the-art, and the typing speed is 40 WPM. MFSK16 was designed specifically for Long Path DX, and also performs well under difficult NVIS [15] conditions because of the ability to handle severe multi-path reception. There is a more sensitive slightly slower version called MFSK8 that has 32 tones and operates at 7.8125 baud.

Morse
There is now good software for transmitting and receiving Morse using a computer, but to date none can match the best human operators. Morse is an ASK chat mode, frequently called 'CW' [16], although this term describes a modulation technique used by several different modes, rather than the mode itself. The use of Morse for radio dates from before the turn of the 20th century.

MT63
An unusual mode best suited to broadcast use, but often used for chat, albeit slowly, MT63 employs 64 phase-modulated carriers keyed at 10 baud. MT63 is another of the 'designer modes', making use of techniques not possible without sophisticated digital-signal processing. Because of the wide bandwidth, long interleaf and special error correction, this mode is very robust and very sensitive. It has a small but enthusiastic DX following on 20m, and because of the easy tuning, suits leisurely chat sessions and nets on the lower bands. It is one

of the best broadcast modes, with 100 WPM throughput, but the 1kHz bandwidth and aggressive nature of the signal makes it unpopular with users of other modes.

PacTOR

A commercial system developed in 1991 as a replacement for AmTOR, for Amateur and commercial service. It has largely (and not before time!) replaced HF packet for message forwarding. PacTOR is a highly effective synchronous connected mode with sophisticated ARQ error correction. The modulation is FSK at 100 or 200 baud, depending on conditions. It is offered in terminal controllers by several manufacturers, under licence from SCS in Germany, and there is at least one DOS software solution. For those interested in message forwarding, the performance and widespread use of the mode for bulletin board use can justify the cost of becoming involved.

Fig 1.4. An excellent SSTV picture received on 20m.

A more recent variant, PacTOR II is backward compatible with PacTOR, but uses a different four-phase PSK [17] modulation system and higher speeds. Since it is only available in high-end commercial controllers, PacTOR II is quite expensive.

PSK31

Probably the most widely known 'designer mode', PSK31 was developed during the late 1990s as a replacement for RTTY, for contesting, DX and conversational QSOs. PSK31 is a differential binary PSK mode that operates at 31.25 baud, and has no error correction. It is an easy-to-use chat mode, and the transmitted signal is only about 50Hz wide! It is great for QRP [18], very sensitive and highly effective on HF, although at times reception is poor due to the effects of ionospheric Doppler modulation. There is a QPSK [19] variant with error correction, but it is not widely used on HF because there is no apparent performance benefit under most conditions.

RTTY

The name stands for Radio TeleTYpe, and the mode has been widely used since the 1950s, at first with mechanical teleprinters retired from Telex service. For much of its long history, RTTY was the only digital mode worth considering, and it underwent a strong revival during the 1980s as microcomputers became available. RTTY is a chat mode with no error correction, uses FSK, and usually operates at 45.45 or 50 baud. It is badly affected by multi-path reception and selective fading. While still widely used, especially for DX and contests, it has been outclassed for conversational QSOs by other modes such as PSK31 and MFSK16.

SSTV

Slow Scan TeleVision (**Fig 1.4**) has not been considered a digital mode for most of its life (it was developed in the 1950s), but since it is now operated almost exclusively by computer, and is so readily accessible, we'll include it here. SSTV is highly popular now that is has finally become inexpensive to operate.

Not normally used as a QSO mode on its own, SSTV is usually used as part of an SSB conversation to pass pictures and add interest to the contact. It operates in many different modes with different speeds and resolution, but all modes use analogue FSK modulation, and most offer full colour pictures.

References

[1] He describes some of these modes in 'New Sound Card Aided Communication Modes', ENZCON Conference Proceedings, 2000.
[2] QRM – interference.
[3] DX – distance (long distance).
[4] Antipodal – related to the antipodes, the exact opposite side of the earth.
[5] QSO – conversation by radio.
[6] WPM – words per minute, the usual method of measuring typing speed. Professional typists exceed 60 WPM. Radio Amateurs normally 'hunt and peck' and rarely exceed 20–30 WPM.
{7] FSK – Frequency Shift Keying (like digital frequency modulation).
[8] Baud – signalling speed, symbols per second.
[9] FEC – Forward Error Correction (information to correct text sent with the text).
[10] ARQ – Automatic Request Repeat (detected errors corrected by asking for repeats).
[11] ASK – Amplitude Shift Keying (like digital amplitude modulation).
[12] DSP – Digital Signal Processing.
[13] MFSK – Multiple Frequency Shift Keying.
[14] FFT – Fast Fourier Transform. A mathematical technique used to transform data from the frequency domain to the time domain.
[15] NVIS – Near Vertical Incidence Signal (HF conditions typical on 80 and 40m).
[16] CW – Continuous Wave, usually implies interrupted continuous wave.
[17] PSK – Phase Shift Keying (like digital phase modulation).
[18] QRP – low power operation (typically 5W or less).
[19] QPSK – Quadrature (four-phase) PSK.

Digital fundamentals

In this chapter
- The digital transmitter
- The digital receiver
- Considering the data
- Data codes

IN THIS CHAPTER we consider what makes up a simple digital-mode transmitter, first using conventional techniques, and then using subcarrier techniques. Then we look at what makes up the digital mode receiver. After this we consider the nature of the data we wish to transmit, and explore ways to improve reception by managing the data.

These fundamental techniques will be described with as little technical complexity as possible, and with no advanced mathematics. Don't worry if you come across names of digital modes you've never heard of, or words you don't understand. They'll all be explained in a chapter devoted to the mode in question, or in the appendices and glossary (Appendix F).

For the technically minded, more detailed descriptions and examples of some of the simpler techniques are also to be found in Appendix A.

The Digital Transmitter

As discussed in the Introduction, and without going into unnecessary detail, transmitting a digital mode involves the modulation of one or more properties of the carrier of a radio transmission.

In simple terms the properties that define a transmission are:

- Amplitude (power or signal strength)
- Frequency
- Phase
- Time

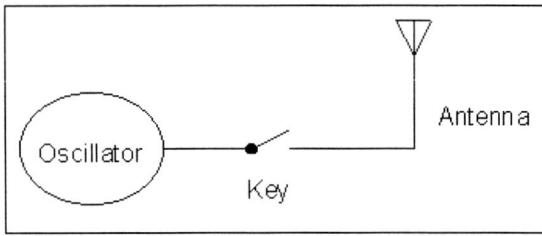

Fig 2.1: The simplest transmitter

To send data, these properties are changed in some way by a device called a modulator. The modulator changes one or more of the first three properties, varying it with time, little different from analogue modes. All digital modes use time to define the duration of each individual item of transmitted data. More about that later.

As a simple example, to transmit Morse, we modulate a carrier in amplitude (on and off), and hopefully do so without have any great effect on the frequency! As in **Fig 2.1**, in the simplest possible amplitude modulated transmitter the key connects the antenna to a signal source.

The conventional way to transmit digital modes is very similar to transmitting Morse, because the modulator, which may be as simple as a switch or relay, operates directly at the RF carrier frequency. For example, to send frequency-shift keyed (FSK modulated) teletype, the carrier frequency of the transmitter is keyed back and forth between two fixed frequencies, achieved by switching a reactive component in the carrier oscillator circuit.

Unfortunately the conventional technique is inconvenient with modern radio equipment, as dedicated and specialized modulators are either impractical, or for some modes, simply not possible. Most Amateurs will wish to use the universally available SSB transceiver.

The Subcarrier Technique

These days, digital mode transmission is the same in principle, but the result is achieved in a different way to that just described, principally in order to allow an SSB transceiver to be used without modification.

This technique, which uses an audio-frequency subcarrier, makes tuning and netting very easy, and also makes changing mode very easy. The subcarrier method also allows the signals to be generated and received by a PC using its sound card. The subcarrier is simply used to create an audio-frequency replica of the intended radio signal. The signal is therefore the same as the signal heard in the receiver.

New digital modes, conventional digital modes, even Morse, and analogue modes such as SSTV are generated in this way. Many of the newer modes can only be transmitted in this way.

Self-contained HF modems, such as are used for packet radio, PACTOR and AMTOR, also use this very convenient technique. In comparing this technique to direct RF modulation, there is no difference in the actual signal transmitted if the equipment is adjusted correctly.

You will recall that the SSB transmitter operates by translating audio frequencies to RF frequencies, by effectively adding (USB) the audio frequency or frequencies to the suppressed carrier frequency (typically the receiver dial frequency), at the same time preserving the phase, frequency and amplitude information of the modulating signal. For LSB the audio frequency or frequencies are subtracted from the suppressed carrier frequency. By generating the signal using a subcarrier, exactly the same sounds are created as you would hear in a receiver tuned to the same frequency.

Block Diagram

Most digital modes have the same basic construction to the 'subcarrier transmitter'. The data to be transmitted, be it from the keyboard or data stored in the computer, is used to key the modulator, which changes the properties of the transmitter subcarrier.

The diagram in **Fig 2.2** is very much simplified. When an SSB transmitter is used, the carrier generator of the conventional technique is replaced with an audio-frequency subcarrier generator.

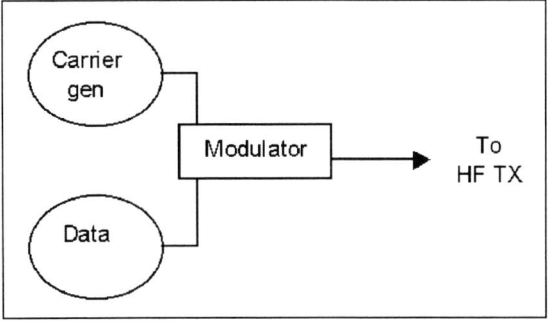

Fig 2.2: Digital transmitter

The data is usually manipulated in several ways, there are often filters before and after the modulator, and there may be more than one modulator. Once it leaves the modulator on its way to the transmitter, the signal is analogue. It is simply an audio signal, and sounds just like the signals tuned in on the receiver.

The modulator details will differ depending on the mode to be transmitted. More than any other part of the system, the modulator and the modulating technique define the character and bandwidth of the transmitted signal.

Sound Card Signals

Perhaps the biggest revolution in digital modes in recent years is the use of the PC sound card. Generating accurate digital signals using a PC sound card is relatively straightforward, but the technique is quite different from traditional hardware analogue or digital techniques. The sound card operates by sending numerical values to a digital-to-analogue (D-A) converter, which it must do at a fixed rate, much higher than the subcarrier frequency. This D-A converter turns the numbers into actual voltages, which put together create audio sounds.

Sine waves are generated by repeatedly sending many values from a table to the sound card at the sampling rate.

A number of other digital signal processing (DSP) techniques are used to generate digital modes. These techniques are highly mathematical, and some have no equivalent in the hardware world, offering important advantages through the use of the PC and sound card. Some of the techniques available have never before been considered possible.

Looking at Digital Signals

Throughout this book the Fast Fourier Transform (FFT) is used as a technique to identify digital modes. **Fig 2.3** is the first of these, a picture called a spectrogram, which displays frequency vertically, time horizontally, and signal strength as brightness. In all cases illustrated in this book the FFT settings are the same [1]. This example is of 20 WPM Morse. The keying sidebands are clearly visible.

The HF Transmitter

In most cases, because of the way the signal is generated, an SSB transmitter is necessary for

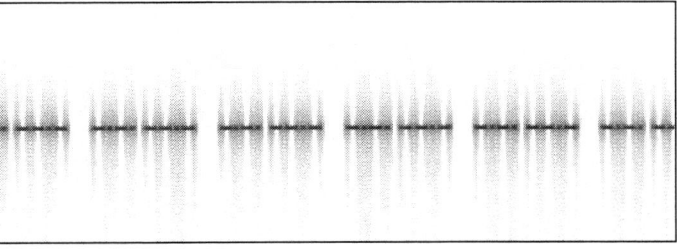

Fig 2.3. Spectrogram (FFT) of Morse

transmission of digital mode signals. The audio subcarrier is fed into the microphone circuit, either via the microphone connector, or via a special accessory socket.

Some modes, notably Morse, RTTY, SSTV and MFSK16, don't require particularly good transmitter linearity, since these modes only ever transmit one carrier at a time. In fact, all these modes can be generated at radio frequencies and used with a conventional class C transmitter.

Other modes which send just one carrier at a time, for example Hellschreiber, use envelope modulation of the signal, which requires reasonable linearity if the signal is to remain as narrow as possible.

Other modes which transmit more than one carrier at a time, or transmit amplitude-modulated carriers, require very good linearity, and the transmitter must be carefully adjusted to ensure that the transmission is clean and free from inter-modulation. PSK31 is an important example of such a mode, and a maladjusted (overdriven) transmitter will make the signal many times wider than it need be, harder to tune, and reception may also be poor.

In most cases, including for the single tone modes, it is easier just to use a linear transmitter for all transmissions – peace of mind, ease of setting up and ease of changing modes come at the cost of a little power output and efficiency.

The Digital Receiver

There is much in common between the digital transmitter and the digital receiver (**Fig 2.4**). However, since it has to deal with less than perfect signals, extra stages are added to the receiver that have an important effect on the performance. The figure below illustrates a representative digital mode receiving system.

Audio from the HF receiver goes first of all into a band-pass filter, which is designed to accept the signal and all its useful sidebands, but reject noise from outside the signal bandwidth. In many cases the signal bandwidth will be rather less than the HF receiver pass band, so this first filter plays an important part in reducing the amount of noise received with the signal.

Fig 2.4: Digital receiver

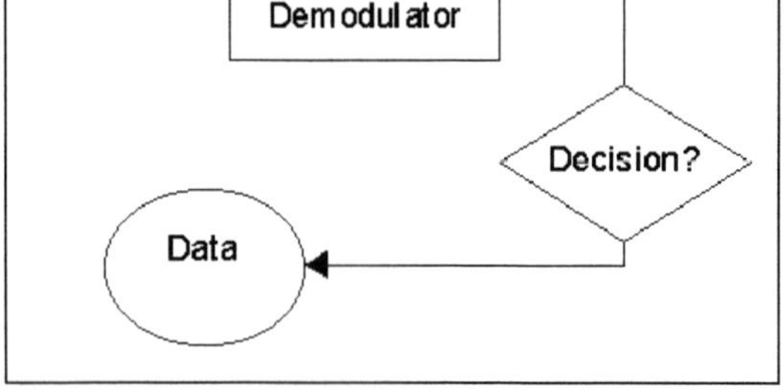

The demodulator has the task of recovering the data from the signal. For ASK, this can be an amplitude detector, just as in an AM receiver; for FSK this is a frequency discriminator of some type, often very similar to an FM detector. For PSK and a number of other modes a phase-sensitive detector or balanced modulator can be used. For some of the newer modes (notably MT63 and MFSK16) an FFT technique is used.

The data from the demodulator may have significant noise present, along with other products introduced in the demodulation process. It will also have some residual subcarrier present. These are largely removed by the low-pass filter.

At this point, the signal is still in analogue form. As well as the filtering in the transmitter and modulation by the ionosphere, the signal is further rounded off by the filters in the HF receiver and the receiving equipment. As a result, the recovered data is nothing like the original signal, and will consist of a rounded and delayed representation of it, along with DC offset, drift and noise. The final stage in the recovery of the original data is the decision process, where an arbitrary signal level is chosen to represent half way between '0' and '1', and the output becomes a square wave. Typically this stage is a comparator, perhaps with hysteresis to reduce the effect of noise.

The decision stage, often called a slicer, usually has some means of compensating for changes in DC offset due to drift, changing signal conditions, and changing data content. Some modes, notably the fuzzy modes, dispense with the decision stage completely, and present the data to the user in an unprocessed state. Since any decision could be in error, especially when reception is poor, this can have important advantages if the ear, eye and brain can be employed to recognise the data received.

Considering the Data

Symbols, Bits and Bauds

The basic element of digital communication is the bit. One binary bit can imply 0 or 1, plus or minus, yes or no, on or off. All letters, numbers and commands to be transmitted are created from these basic elements, by using different coding schemes. The rate at which these bits can be transmitted down a wire is a fundamental property of the transmission system, and is measured in bits per second or bps.

Just to complicate the issue, the basic signalling entity of a digital radio transmission, the symbol, may in some cases imply more than one bit – for example a QPSK transmission, where the phase can have any one of four values and carries two bits of information per symbol, so the bit rate is twice the symbol rate.

When considering the bandwidth of the signal, and the effect of the ionosphere on a radio transmission, it is the speed at which each symbol is transmitted that has to be considered, not the speed of each bit. The unit of measurement of these symbols, the number of symbols per second, is the baud, in recognition of Jean-Maurice-Émile Baudot, developer of early telegraph transmission techniques.

However the user most likely wants to know about the data signalling rate, in bits per second, so it is important to know the relationship between symbol rate and the raw data rate for the mode being considered. **Table 2.1** gives some examples.

Bits, Bytes and Characters

The fundamental unit of text transmission is the alphabet character. These are coded in such a way that a group of bits represents each letter or number. Hence the speed of text transmission depends on how many bits are required to define each character, as well as the bit rate.

As an example, asynchronous RTTY (radio-teletype) uses a system with five data bits and 2.5 synchronising bits per character, so each character requires 7.5 bits to transmit [2]. At a speed of 45.45bps, that amounts to 45.45 / 7.5 = 6.06 CPS (characters per second).

Table 2.1 Typical Amateur Mode Data Rates

Mode	Symbol Rate (baud)	Raw data rate (bps)	Typing Speed (WPM)
RTTY	45.45	45.45	60
PSK31	31.25	31.25	~35
MFSK16	15.625	62.5	~40
MT63	10	640	100

The average English language word contains about six characters (including the following word space), so 6.06 characters per second represents about 60 words per minute (WPM). Compare the RTTY performance with other modes in the 'Typing Speed' column in Table 2.1.

RTTY and PSK31 do not use error correction. The other modes in the table use error correction, so the relationship between the raw data rate [3] and the typing speed is not immediately obvious.

Other modes generally use more bits per character than RTTY. Some modes even have a variable number of bits per character – and a good example is Morse, where frequently used letters like 'E' and 'T' have fewer elements than lesser used ones such as 'J', 'Q' and 'X'.

There are two widely recognised alphabets used for digital modes. These are the International Telegraph Alphabet No 2 (ITA2) or 'Murray' alphabet (used for RTTY, for example) and the International Telegraph Alphabet No 5 (ITA5) or ASCII Alphabet, used for packet radio, MT63 and many other modes.

ITA2

ITA2 is widely misnamed as the 'Baudot code', but it was in fact developed by Donald Murray. Apart from the synchronising aspects, the ITA2 alphabet is essentially the same as that used by Murray, a New Zealand journalist and inventor who worked in Paris, and who made important contributions to remote typesetting for newspapers from 1903 [4].

ITA2 uses only five data bits per character for transmission, so has only 2^5 or 32 possible combinations. In order to define a useable set of characters, the alphabet is divided into 'shifts', consisting of two otherwise ambiguous character sets, 'letters' and 'figures'. Two of the alphabet characters are dedicated to switching between these sets. The concept of shifts is very old – it was used by Emile Baudot, who used the idea with his five-key manual keyboard.

The letters shift contains all upper case letters, and the control characters 'LTRS', 'FIGS', 'SPACE', 'LF' (line feed) and 'CR' (carriage return). One combination (all zeros) is generally unused as it was used in the past for punching out errors in paper tapes prepared for transmission.

The figures shift contains all the numbers 0–9, a limited range of punctuation and the same control characters as letters shift. The punctuation tends to differ widely from one country to another [5].

ITA5

Better known as 'ASCII' (American Standard Code for Information Interchange), this alphabet arose from the fledgling computer industry, and is the internal alphabet used in every personal computer used today. Since it uses seven data bits, many more characters can be defined than in ITA2, and no notion of 'shift' is necessary. The 2^7 or 128 combinations are also organized in a regular manner, so the number and letters follow in a logical order. Letters have upper-case and lower-case versions, and there are many control functions defined as well [6]. The 'extended ASCII' character set has 256 combinations, but the 128 extensions are not standardized, being strongly dependent on language and computer system.

Varicodes

Some modes, for example Morse, PSK31 and MFSK16, use alphabets which do not have a fixed number of bits per character. This idea has two major advantages:

♦ More frequently used characters can be assigned shorter sequences and therefore be sent more quickly.

♦ The size of the alphabet is not limited by the number of bits. For example it is possible to increase the alphabet to several hundred characters, making transmission of Japanese and other pictogram- based languages practical.

The varicodes are normally designed to transmit average text at maximum speed, so the assignment of codes is decided by statistical analysis of text. For example, it is believed that Morse code assignments were made on the basis of the amount of type in a printer's type trays! This does mean that some other text, for example in other languages that use the same character set, will probably not demonstrate the same improvement. In most cases the performance difference is trivial.

Varicodes are not used with asynchronous techniques, as by their nature asynchronous systems must have a regular starting point and ending point for each character to be determined.

Character Synchronisation

Whether each character consists of a fixed number of bits or not, some method is necessary to decide where each character starts and stops. This is called character synchronisation (sync for short). The point at which one character stops and the next starts needs to be marked by a unique event, a unique combination of bits, the sync bits.

Asynchronous (async) systems like RTTY identify the start of each character with at least one and a half bit-times of '0' followed by one bit-time of '1'. This technique is simple enough, and ideal for mechanical teleprinter machines, but is not totally reliable, because if synchronism is lost due to noise, characters containing the sequence '001' may be misunderstood as the start of a character. This sort of error may cause garbled reception for several characters until the correct synchronism is recovered.

An asynchronous system is able to send characters at varying rates. At the end of each character, the terminating '0' can be any length from 1.5 bit-times to infinity, and is called the 'stop bit'. When a new character is sent, the stop bit is terminated and the '1' or 'start bit' is sent, followed by the data [7].

Synchronous systems were developed to overcome this loss of synchronism, but at the expense of greater complexity. One synchronous transmission type, which we'll call 'block transmission', requires repeated transmission of a special synchronising bit-combination at the start of transmission, to allow the receiver to find the correct starting point, and may repeat the sequence occasionally throughout the transmission [8].

Another type, which we'll call 'bit-stream transmission' uses a unique sequence of bits to terminate each character, and this sequence is not permitted within the character. Systems of this type can re-synchronise anywhere in the transmitted message [9].

Synchronous systems cannot handle pauses in the middle of a transmission, as can happen in asynchronous modes such as RTTY. This is because the receiver must continue to accurately time and count each bit. The solution in a block transmission system is to terminate the transmission, wait, and start again later with a new block. In a bit-stream transmission, filler characters that mean nothing (but keep the receiver happy) are sent until new data arrives.

Data Codes

So far we have dealt with simple methods of encoding characters, bytes or other identifiable lumps of data in order that they may be transmitted efficiently and identified correctly on receipt. There are four additional types of coding which can be used to manipulate the data. These are:

- Codes used to manage statistical data content
- Codes used to provide error-correction information
- Codes used to spread the signal to achieve diversity
- .Techniques used to obscure the transmitted data and prevent unauthorised reception

Baseband Codes

These codes are used to ensure that, for example, the transmission does not include long sequences of '0' or '1' which might confuse the receiver; or to keep the number of '0's and '1's about the same, to maintain DC balance in AC coupled circuits. [10]

Codes of this type can also be used to manage the bandwidth of the transmission and to provide signals that can be received correctly even when inverted (i.e. when '0' is received as '1' and vice versa. [11]

Error Codes

Error coding adds extra information in order that errors in the received data can be detected, and very often corrected without the data being retransmitted. This type of reconstructive error correction is called Forward Error Correction (FEC). If errors are detected and re-transmission is required for the data to be corrected, the technique is known as Automatic ReQuest repeat (ARQ). The various Error Coding Correction (ECC) techniques are covered in more detail in Chapter 3.

References

[1] The settings used throughout the book are 5.5kHz sample rate, FFT size 1024, Scroll mode (BW), dwell 2 ms, spectrum average = 1. The software is Spectrogram V 4.2.6.4.

[2] See Chapter 3 for details.
[3] The data rate including any error-correction information.
[4] The alphabet Baudot developed some three decades earlier (1874) was used with a five-key manual sender rather like a stenographer's keyboard. It bore no resemblance to ITA2, and was not used for radio transmission. It became ITA1 and was used by some early teleprinters.
[5] See Appendix B for details.
[6] See Appendix B for details.
[7] For more complete details see Chapter 14.
[8] A technique frequently used in ARQ modes. Packet radio, PACTOR and CLOVER use this technique.
[9] Frequently used with Varicode alphabets, as in PSK31 and MFSK16.
[10] HDLC, used in packet radio, is this type of code.
[11] Manchester codes have this property; so do the differential coding techniques used for PSK.

3

Serial data transmission

In this chapter
- Data synchronism
- Error correction techniques

UNDER MOST circumstances, transmission of data via a radio channel or on a single wire allows only one item of digital data to be sent at a time, ie the radio channel can only be in one of two states (on/off, high/low, one phase or another).

Thus in most circumstances we need to deal with data sent one bit at a time, ie serially. So, if the data sent is for example seven bit ASCII text, each bit needs to be sent one after another, one at a time.

While some modes do transmit more than one bit at a time, the data is very frequently disassembled and reassembled as a stream of serial bits anyway. Thus, it is pertinent to look at simple serial-transmission techniques in some detail, as the techniques apply generally. Chapter 2 described data bits, symbols and alphabets, and in this chapter we see how these are strung together and end up on the air.

The user does not need to know in detail how serial transmission works, but it is helpful to understand what goes on inside an MCU such as the Kantronics KAM (**Fig 3.1**) or inside the PC software, in order to make best use of the technology.

Fig 3.1. A popular MCU for FSK serial modes (photo: Ted Minchin)

Data Synchronism

It is easy to imagine sending data bits one at a time, but how is the receiving equipment to know when each bit starts, and when it finishes?

25

Further, how is the receiving equipment to know which bits belong to one character, and which to the next? There are thus three problems to solve:

- Correct bit rate (speed or clock rate)
- Correct bit synchronism (ie where the centre of each transmitted bit is)
- Correct character synchronism

If the bit rate or clock rate at the receiver is slightly different to that at the transmitter, the data sampling point will drift as further data is received, and errors are inevitable.

As a general rule, the best place for the receiver to sample the incoming data bits is around the centre of each bit, hence the interest in locating this point accurately. The further away from the centre of the bit that the data is sampled, the higher the risk of error. This is because (a) the received signal timing varies as it passes through the ionosphere, and (b) the edges of the data bits become very rounded and tend to be noisy.

The character synchronism problem is one of how to recognise the end of each group of data bits (say those of one character) and the beginning of the next. Very often the character synchronism and bit synchronism are performed by a common mechanism. When a digital mode is designed specifically for HF use, much attention is paid to the choice of a synchronising technique for best reliability. There are numerous effective methods used for achieving bit and character synchronism [1], of which the following are the most common:

- Character asynchronous
- Character synchronous
- Bit-stream synchronous
- Block synchronous
- Quasi-synchronous
- Non-synchronous

The last two methods don't involve synchronism at all, and yet are surprisingly effective.

Error Correction Techniques

Mechanised data transmission and reception was first developed over 100 years ago, mostly with a view to increasing transmission speeds and to allow messages to be printed on paper or tape, but another big advantage offered was the reduction of operator-induced errors.

Nothing further was done for nearly 100 years to reduce errors, except through use of repeated phrases, multiple transmissions and multiple reception (repetition and diversity) to ensure that the message got through. This was generally adequate on wire lines with low noise levels, but not sufficient for noisy radio channels. Noise, varying transmission path properties and interference can have very serious effects on data transmission.

Modern error correction techniques date from 1948, when Claude Shannon [2] postulated that by adding redundancy (extra data) to a transmission, the data could potentially be recovered at very low signal-to-noise ratios. In effect, transmission speed could be traded for both sensitivity and reduced error rate [3].

Practical examples of error reduction resulted from the work of numerous famous mathematicians, but the theory behind this work is well beyond the average amateur's 'need to know'! Instead, the various broad categories and techniques will be described. ECC is used in many different types of Amateur modes; for example QPSK31, MFSK16, MT63, PacTOR, Clover, Q15X25, AX25 packet, and AmTOR. High redundancy without error coding is also used by the 'fuzzy' modes.

Forward Error Correction

Forward Error Correction (FEC) coding adds extra information as suggested by Shannon, in order that errors in the received data can be detected, and correct data reconstructed and very often corrected without the data being retransmitted. FEC is an important technique used when repeating the data on request is not practical – for example a one-way transmission (a broadcast, net operation or data from an inter-planetary probe), or when the data has a high requirement for timeliness, such as data from a music CD.

AmTOR mode B is a practical example of an FEC mode used for broadcast purposes (eg used for calling CQ). FEC is also useful for 'chat' modes, allowing several people to receive reliable signals at the same time.

ARQ

If errors are simply detected, and retransmission is required for the data to be corrected, the technique is known as Automatic ReQuest repeat (ARQ). AX25 (packet radio) and AmTOR mode A are modes of this type. AX25 [4] uses a CRC [5] algorithm to send and detect errors, while AmTOR uses a simple parity [6] technique. A further advantage of the ARQ technique is that it is possible to use the error performance information to manage the modulation properties of the link, in order to maintain maximum possible throughput under varying conditions. ARQ is not practical where one station transmits to more than one other station.

Multiple Techniques

Some modes employ both FEC and ARQ options. Modes such as Clover, Q15X25 and other newer packet protocols use FEC to correct most errors, and resort to an alternative ARQ technique only when an error cannot be corrected by FEC. The efficiency is improved because less time is lost on repeats which involve changing the link direction.

PacTOR uses a different technique – a CRC is transmitted with each block of data, and by solving the CRC over repeated ARQ cycles, the data can be assumed to be correct. However, the blocks containing errors are not abandoned, since they may contain mostly correct data and just a few errors. The FEC technique repetitively averages the blocks received in error with previous ARQ attempts until the averaged CRC of these attempts indicates that the averaged data is correct. This method is called 'memory ARQ', and is very effective in conditions with burst noise, as the complete data block can be reconstructed without ever receiving the whole block correctly.

There are further complex coding examples where multiple coding techniques are used, one after another. The very weak signals received from interplanetary probes are invariably of this type.

Error Coding Methods

There are three main types of Error Correction Coding (ECC) used in Amateur modes:

* Block codes
* Systematic codes
* Convolutional codes

Connected modes usually use a block coding technique, where complete blocks of data are processed prior to and following transmission. Extra error correction information is then added to the block after the data in the manner of a CRC, check sum [7] or parity. AX25 (packet radio) Clover and PacTOR use this technique, which is best suited to ARQ operation.

Chat and broadcast modes tend to use linear codes, where the ECC data is coded bit-by-bit continuously throughout the transmission. The simplest of these linear codes is the 'systematic' or parity code, based on a simple mathematical or logical rule [8].

By far the most common (and most powerful) technique for bit-stream transmissions is the 'binary convolutional coder' where a mathematical process generates extra information describing the past history of the transmitted data, enabling the correct data to be fully reconstructed [9].

Phil Karn KA9Q [10] has provided numerous coding suggestions and working software examples for Amateur use. His FEC algorithms are for example used in MFSK16.

Use of High Redundancy

An important and very simple alternative to error coding is to use redundant transmissions. For example, AmTOR mode B (FEC) transmits every block of data twice. The military ALE system transmits each block three times.

Hellschreiber could be considered highly redundant, as it takes 49 data bits to define a character that can be transmitted by RTTY in a mere 7.5 bits. However, the Hellschreiber character will still be recognisable as the same character (and never as another) with perhaps 50% of its data bits received in error, while the RTTY character will be lost or changed to something quite different with just one bit error (14%).

Fig 3.2 Illustrates the ability of the eye and brain to recognise redundant text with a high error rate. While the text may be difficult to read, there is no possibility of misreading the letters or words as any other.

Fig 3.2. Redundant text in noise

Interleavers and Spreading Codes

When ECC is used, it is important to give the decoding system the best possible chance of producing the correct result. Some coding techniques cope with a high level of random errors, but are especially prone to burst noise (such as lightning), where several data bits close together are corrupted. In order to reduce the effect of these burst errors, a technique called interleaving is used, where the data bits are muddled up and transmitted in a different (but known) order. The receiver restores the data order before decoding it. Then, when a burst of noise causes loss, the noisy or lost bits are spread through the message and do not easily disrupt the decoder. The interleaver spreads the signal in the time domain, but except perhaps for synchronisation of the interleaver, no additional data need be added [11].

It may seem perverse to add further data to a transmission in order to make the signal wider, but this technique is frequently used to make the signal more robust [12]. It is simply another way of adding redundancy, like a frequency-domain equivalent to interleaving. Spreading works because when the data is 'unspread' at the receiver, the noise bandwidth is reduced. The noise and errors received with the data are spread uniformly throughout the data and so can be removed more readily than when the errors occur together. Spreading is a very powerful technique to combat carrier interference and other narrow-band interference.

When a long random-number bit sequence called a 'maximal length sequence' is used to spread the signal, in addition to making the signal uniquely recognizable, very powerful correlation techniques can be used to improve receiver sensitivity. The technique can also gives precision timing data for accurate synchronisation even when the signal is very weak, and can allow accurate measurement of propagation delays between transmitter and receiver.

Encryption Techniques

Cyphers [13] are codes with hidden meanings. Radio Amateurs are not permitted to use these encryption techniques on the air. It is acceptable to use universally known abbreviations (eg the 'Q' codes), or to transmit by packet radio files compressed using a public compression algorithm. It is a small step from there to transmission of locked compressed files or encrypted files which would not be permitted. E-mail encryption techniques are also not permitted. Thus the digital-mode Amateur needs to be careful.

It is important to recognise that all the other coding techniques described in this book are devised to facilitate transmission and reception, and all use universally recognised and publicly-available alphabets, codes, code tables and algorithms.

Encryption is a fascinating subject, with a long and interesting history, and mathematically has much in common with signal processing and error coding. Unfortunately it has no place in Amateur Radio.

References

[1] See Appendix A for a description of each method.
[2] Shannon, C. 'A Mathematical Theory of Communications.' *Bell System Technical Journal*, July 1948.
[3] Shannon, C. 'Communication in the Presence of Noise', *Proc IRE*, 1949.
[4] AX25 – X25 packet switching protocol applied to Amateur Radio.

[5] CRC – Cyclic Redundancy Check. A complex polynomial calculation is used to derive a unique number from all the data in the message, and this number is transmitted with the message. When the calculation is repeated at the receiver, the data can be assumed to be completely correct if the calculation matches the transmitted CRC. This technique is a much stronger and more reliable way of detecting errors than the use of parity or a check sum.

[6] A very simple error-checking technique, in which an extra bit or bits are added to a character in order to maintain a constant ratio of '0' and '1' data bits. Obviously one bit in error per character will be detected, but two may not.

[7] A simple logical calculation made on all the data in a message results in a check sum which can be used at the receiver to check the validity of the data. The check sum is easily fooled if there are two errors in the same data.

[8] AmTOR uses this type of coding, and the 7-bit Moore parity code.

[9] This is well beyond the scope of this book. For an excellent tutorial, see http://pweb.netcom.com/~chip.f/Viterbi.html .

[10] See http://people.qualcomm.com/karn/code/fec/

[11] Interleavers are used in MFSK16 and MT63.

[12] GPS satellites, military transmissions like STANAG 4285, and the MT63 mode use this technique.

[13] In the encryption business, a cypher is a letter-substitution encryption, while a code is a word substitution encryption. In this book code has another meaning, namely a publicly and universally known numerical representation of an alphabet character, so the word cypher is used to encompass hidden letters or words in the whole range of encryption techniques.

4

Digital modes and the ionosphere

In this chapter
- Noises and interference
- Multipath reception
- Doppler modulation effects
- Comparative performance
- Designing modes for HF performance

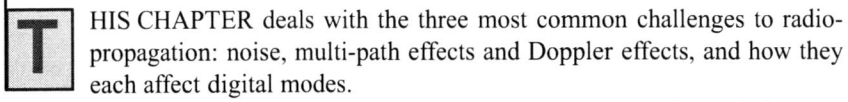

THIS CHAPTER deals with the three most common challenges to radio-propagation: noise, multi-path effects and Doppler effects, and how they each affect digital modes.

As with other modes, digital modes are affected by the properties and phenomena of the propagation medium and by reception conditions. In most situations this means the ionosphere [1], although there are effects caused locally, in the troposphere [2] and in the stratosphere [3]. While operators of other modes are well advised to study these phenomena, and understand how best to take advantage of them, or avoid them as the circumstances dictate, for the digi-mode operator, understanding these propagation phenomena is fundamental to achieving good results.

> **A good understanding of HF propagation phenomena is fundamental to achieving good results on digital modes.**

Noise and Interference

While noise and interference can be both natural and man-made, it is best to consider noise from both sources in three broad categories: broad-band noise, impulse noise, and carrier interference.

Broad-band Noise

There are two main causes of natural broad-band noise, and any number of man-made sources. The sun directly generates noise, radio energy radiated from active areas of the sun and particularly strong in the UHF and microwave regions.

The sun also indirectly causes auroral noise, where radio signals are noise modulated by highly active charged particles in the ionosphere.

Thunderstorms have highly active clouds with strong internal air convection currents which generate electricity through particle movement and friction. In addition to causing lightning, these clouds can cause very strong noise. When a charged cloud passes low overhead, charge is attracted by the clouds, and is sprayed into the air from antennas and other grounded objects, generating very strong noise in receivers. Fortunately this effect is not common, and it's wise to avoid operating when it happens!

There are many man-made sources of wide band noise; for example, high voltage discharges from power lines, TV receivers, and welding equipment. Many of these sources are also strongly modulated at AC supply frequencies. The natural limitations of receiving equipment generates noise, especially at UHF and above, and although generally HF signals are well above the noise level, fading conditions, weak signals, physically small antennas and higher-frequency operation can result in reception limited by receiver noise.

The digital mode PSK31 is widely known for its sensitivity. In part, this is because the PSK31 signal can be copied well below the noise level, enabling communications to take place over great distances with low power. MFSK, MT63, PacTOR II and PSK-Hell are among other modes with excellent sensitivity. Narrow-band modes generally give the best broad-band noise performance.

The brief examples in **Table 4.1** illustrate the receive performance of popular modes under broad-band noise conditions. The signals were generated by accurately mixing a locally generated audio signal with off-air noise, with the signal-to-noise ratio set at −3dB, ie the signal was weaker than the noise by 3dB. The noise–power bandwidth was 2.4kHz.

There would seem from this to be little difference between some modes as far as performance goes, but of course broad-band noise performance is only part of the story!

Impulse Noise

The most common cause of impulse noise is lightning. The effects are random in nature, but so immense that during stronger lightning strikes and for a significant period afterward, the receiver is completely overloaded and desensitised. Automotive ignition noise, electric fences, various ionospheric sounders and 'woodpecker' type radar transmissions can generate similar noise, although generally more regular and fortunately not usually as strong. Morse key clicks and strong SSB splatter can cause similar effects.

For voice communications and human-read Morse, gaps in reception due to noise bursts can be filled in by the brain, from context, or by requesting repeats. In addition, good noise blankers can help reduce the overload effects, allowing the receiver to recover faster with less desensitisation.

Unfortunately digital modes tend not to handle impulse noise well, as the loss of as little as one bit every now and again can be enough to render reception unintelligible.

Digital transmissions generally use modulation techniques, such as FSK and PSK, that are moderately insensitive to amplitude variations, and therefore to

CHAPTER 4: DIGITAL MODES AND THE IONOSPHERE

impulse noise. It is also important to employ a mode that has the ability to hold synchronism well.

However, as mentioned above, the receiver is also sensitive to impulse noise because it can be overloaded or at least temporarily desensitised by the impulses. Obviously the receiver must recover quickly from the impulse to minimise the number of data bits received in error.

For impulse noise to have the least possible effect on a digital transmission, the transmission must also contain sufficient redundancy for the information to be recognised, or recovered, with 'holes' chopped out of the data by the noise. There are two excellent ways to employ this concept – either by using high redundancy and a receiving technique where synchronism is not upset by noise (e.g. Hellschreiber modes), or by using an error correction technique to compensate for and correct errors caused by gaps in the data.

Error correction with good impulse noise rejection generally uses a data interleaving technique to ensure that the impulse-noise-induced errors are spread evenly to maximise the error-correction performance. Both of these techniques were discussed in Chapter 3.

Lightning-induced impulse noise is of course worst on the lower HF bands, especially at night. These bands also suffer from man-made electrical noise at any time. Ignition noise generally affects the higher bands and VHF.

Table 4.1. White noise performance.

RTTY
THE QUICK BRNWN FOG JUMPS OMT TNZ TATY DOH88$678.,,

BPSK31
THE QUICK BROWN FOX JUMPS OVER THE LAZY DOG 1234567890

QPSK31
THE QUICK BROWN FOX JUMPS OVER THE LAZY DOG 1234567890

MFSK16
THE QUICK Bcef5 FOX JUMPS OVER THE LAZY DOG 123*d67890

MFSK8
THE QUICK BROWN FOX JUMPS OVER THE LAZY DOG 1234567890

MT63 (1K)
THE QUICK BROWN FOX JUMPS OVER THE LAZY DOG 1234567890

PSK-HELL

FELD-HELL

FELD-HELL 1/8

Simulations of static are by their very nature difficult to generate repeatably, and difficult to measure, as static is highly variable, and measurement depends on the weighting of the meter used [4]. In the examples in **Table 4.2**, mild received HF static in a 2.4kHz bandwidth (rather stronger than the signal, but at about the same average level for each example) has been superimposed on locally generated signals.

MT63 and Hellschreiber (especially PSK-Hell) are particularly effective at rejecting impulse noise, and consequently permit operation at low power even under the noisiest conditions.

Receivers for quasi-synchronous modes (Hellschreiber), and synchronous modes (PSK31, MT63, MFSK16) handle impulse noise better than asynchronous modes (RTTY, computer-read Morse, SSTV). The author has logged many contacts on the 80m band, which is notoriously noisy, using Hellschreiber and MFSK16 at only 5W, including contacts with several stations over 3000km away.

Carrier Interference

On HF, especially 160m, 80m and 40m, we hear broad noises that are easily mistaken for random noise, but seem to have a buzz or hum quality. They can cause the same problems as broad-band noise, but also contain strong spectral

---

Table 4.2. Impulse noise performance.

RTTY
HZ QUICK BROWN FOV JUMPS OMER THE LAZYNDOG 11345678QKV

BPSK31
TPE _UICK BROWN FOX ä.MPE to7ER THE LAZY DOG 12x45670ee

QPSK31
THE Q CK BROW r FOyniMPS±itER THE LAZ 1 tOtt 12Tos567890

MFSK16
THE QUICK BROWN FOXi eTec=kTHE LAZY DOG 12345678xr

MFSK8
THE QUICK BROWN FOX JUMPS OVER THE LAZY DOG 1234567890

MT63 (1K)
THE QUICK BROWN FOX JUMPS OVER THE LAZY DOG 1234567890

PSK-HELL

FELD-HELL

FELD-HELL 1/8

CHAPTER 4: DIGITAL MODES AND THE IONOSPHERE

carrier components. Electricity-supply buzz (**Fig 4.7**) is one example. It is commonly generated by non-linear devices such as rectifiers, triacs and ignitrons – light dimmers and microwave ovens are common culprits. It can also be generated by insulation breakdown, as in dirty high-voltage insulators, and by a myriad of fault conditions in the power network or home appliances.

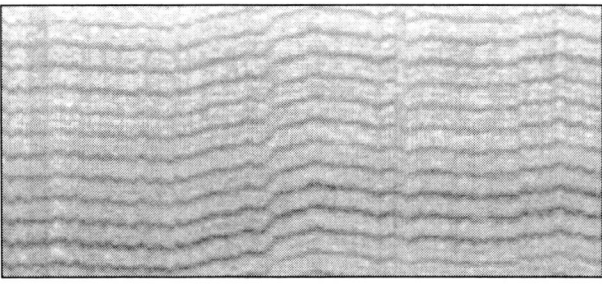

Fig 4.7: Mains buzz has strong 50 or 60Hz components

The tell-tale difference between this type of noise and broad-band noise is that it has a 'tunable' quality – it sounds slightly different as you change the receiver tuning, and also changes in nature with time as the mains frequency varies slightly. Using a spectrogram, it is easy to identify the 50 or 60Hz components in the noise.

Television and computer-clock-frequency harmonics are very similar to mains buzz, but are more widely spaced (**Fig 4.8**). You will be able to find TV harmonics throughout the HF spectrum if you look carefully. They are constant in frequency (often extremely accurate) and can vary in amplitude with picture content, will be spaced apart by the TV line frequency (15.625 or 15.75kHz) and have strong 50 or 60Hz sidebands. They can be a serious problem on LF and the lower HF bands.

Look carefully at 3750kHz (3780kHz in America) – there will be a harmonic exactly on frequency, and it may be extremely precise, dependent on the timing reference used by the local TV network. The noise you hear is a sum of the emissions of all the local TV receivers.

Computer noises tend to move about in time with computer activity. Sometimes the processor, keyboard or printer can be the culprit, and these usually have rhythmic variations, but the most likely source of this type of interference is the computer monitor. Computer monitor noises are very like TV interference, and will look the same on a spectrogram. A good way to check whether the monitor is the noise source is to turn the monitor off momentarily. Sometimes a change in monitor resolution setting will change the frequency of the noise and allow the signal underneath to be received.

While these buzz noises are a type of carrier interference with closely-spaced carriers, the type that give the most trouble to digital modes are genuine transmitted carriers; Morse transmissions, RTTY and other digital modes, short-wave broadcasting and commercial services, and the many unexplained carriers that we hear. Be aware that some of these interfering signals may be generated by equipment in the shack!

Some digital modes are reasonably immune to carrier interference. For example PSK31 and PSK-Hell are not affected by even strong carriers unless they are close to the signal-centre frequency (within about 60Hz or 200Hz respectively). The receiver however

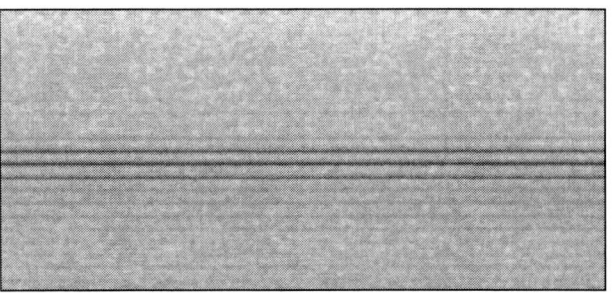

Fig 4.8: TV interference has strong 15kHz components with 50 or 60Hz sidebands

35

will most likely be desensitised by the interference, and for this reason it is always a good idea to use a steep-sided narrow IF filter in the receiver to keep the interference and adjacent signals out. An audio filter or audio DSP filter will not help in this respect.

MFSK16, RTTY and other FSK modes (AmTOR, PacTOR) are not affected by carriers outside the band of the transmitted tones, but are stopped dead by a carrier stronger than the received signal if it is within the signal bandwidth. However, MFSK16 and RTTY are not seriously damaged by most of the burst and broad-band noises.

MT63 and other broad-band modes [5] have excellent carrier interference rejection. MT63 can cope with a continuous strong carrier, Morse or even RTTY interference within the signal bandwidth with no visible effect on reception. It is known that effective MFSK16 and MT63 communications can take place on the same frequency at the same time without serious mutual interference!

Multi-path Reception

Multi-path reception occurs when the signal from a distant station arrives via two or more different propagation mechanisms, for example ground wave and single-hop sky wave on HF, or line-of-sight and refraction or reflection on VHF or UHF.

Because the paths have different properties, the received signals behave differently. In addition, because the different paths have different transit times, the signals can differ in phase and arrival time, and of course interact when they arrive in the receiver.

Very deep fades occur, followed by strong enhancement as the signals add and subtract. At other times during the fading process, one or other of the signals may predominate. While the vast change in signal strength may constitute a problem, for digital modes there are even worse consequences of multi-path reception.

Selective Fading

When the frequency difference between the signals from different paths is very small (less than a few Hertz), a phenomenon called 'selective fading' occurs. This effect is commonly noticed on HF commercial broadcast stations as a 'whooshing' sound, accompanied by apparent changes in the audio-frequency response and severe distortion as the carrier passes into the selective-fading zone. This problem is more prevalent than we generally realise, and can severely affect digital modes.

Some parts of signals are lost in the very deep fade, which may be perhaps 50–100Hz wide, while other parts are received full strength. MT63 is a mode which handles these problems very well, and it also serves to illustrate very graphically the effect of selective fading.

Fig 4.9 is a spectrogram. The horizontal axis represents time (about 15 seconds), while the vertical axis represents frequency.

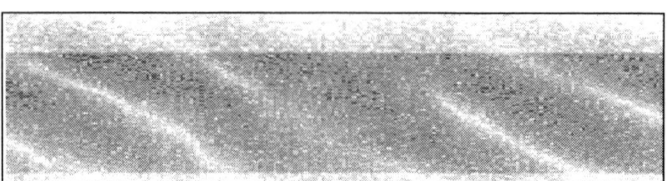

Fig 4.9: An MT63 spectrogram from 20m

The width of the spectrogram is a little under 2kHz, while the signal is the darker ribbon across the spectrogram, exactly 1kHz wide. The darkness of the spectrogram represents signal strength.

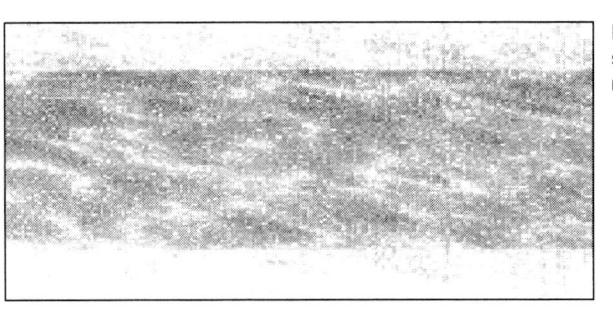

Fig 4.10: 80m selective fading is more complex

Notice the white bands passing diagonally through the signal. These are selective fades over a 2500km night-time path on 20m. There are probably only two F-layer paths involved in this reception example. The nature of the fades can be much more complex if more paths exist. For example, **Fig 4.10** was received from the same station on the same evening, but on 80m.

Very long distance signals exhibit similarly confused selective fading properties. High performance commercial and military digital modes frequently include channel measurement techniques which allow for continual compensation of path-loss variations across the receiver bandwidth.

Timing Effects

With variation of arrival times over different paths, it is inevitable that the internal timing of the received signals will vary. This has little noticeable effect on voice transmissions, and serves only to restrict the receiving speed for Morse, but can have a serious effect on digital modes.

Consider first so called NVIS (Near-Vertical Incidence Signals) operation (**Fig 4.11**), where the signal passes up into the ionosphere at a very high angle, slows down and is reflected back to earth, where at the receiving antenna it interacts with the ground-wave signal.

The delays occasioned by high-angle reflections can be very long, not because of the altitude from which they reflect, but because the refractive index of the ionosphere at that altitude causes the velocity of the radio waves to slow down markedly in the reflective region. Signals with high incident angle penetrate further into the refractive layer than grazing-angle signals and are therefore slowed much more.

Fig 4.11: The NVIS propagation mechanism

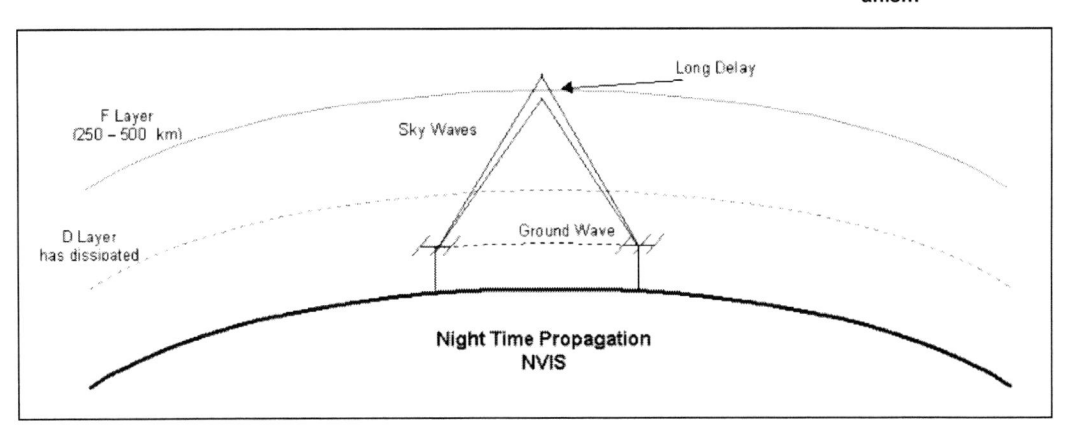

Fig 4.12: Measurement of sky-wave delay

It is relatively easy to measure this NVIS delay, which can be as great as 20ms, and is typically 5ms or more. In **Fig 4.12**, a Feld-Hell signal is received on 3.5MHz over a 50km path, with a ghosting effect above the characters. Each column of a Hell character takes 56ms to send or receive, and each column contains 14 sampled pixels, 4ms each. By magnifying the picture to view and count the pixels, we can measure the ghosting or echo effect caused by the sky-wave path to be two to three pixels, or about 8 to 12ms.

The ground wave delay over this path is a mere 166μs, while the travel time at the speed of light via the ionosphere to an altitude of say 300km might be expected to produce an additional path delay of 1ms, much less than the actual delay shown in Fig 4.12.

Consider the effect this delay has on reception of digital modes with transmitted symbols about 20ms long (for example RTTY). From time to time the sky wave will dominate, and the signal will be delayed by nearly half of a bit-time, and at other times the ground wave will predominate and the timing will change again.

Under these conditions signal distortion is considerable and reception will essentially be non-existent. With faster modes the problem is even worse. FAX and SSTV reception is also poor. Experienced RTTY operators will recognise this problem – 80m operation at night is generally impossible, no matter how strong and apparently clear the signal is. This is the time to deploy specialized modes with good NVIS performance [6].

HF DX operation (**Fig 4.13**) relies on reflection from the ionosphere, typically the F-layer or layers, and from the ground. In many cases two or more paths are viable, and the signals fade up and down so that delays vary, typically by as much as 10ms. This type of fading is relatively slow, but the magnitude of the delay variation can cause loss of data and synchronism.

During the daytime the F-region splits into two layers, so there exist numerous different reflection mechanisms, each with different delays. Of course the modes with fewest reflections (and shortest delays) will predominate since they exhibit least attenuation.

Fig 4.13: The HF DX propagation mechanism

On night-time paths (**Fig 4.14**), when the ionosphere is no longer bombarded

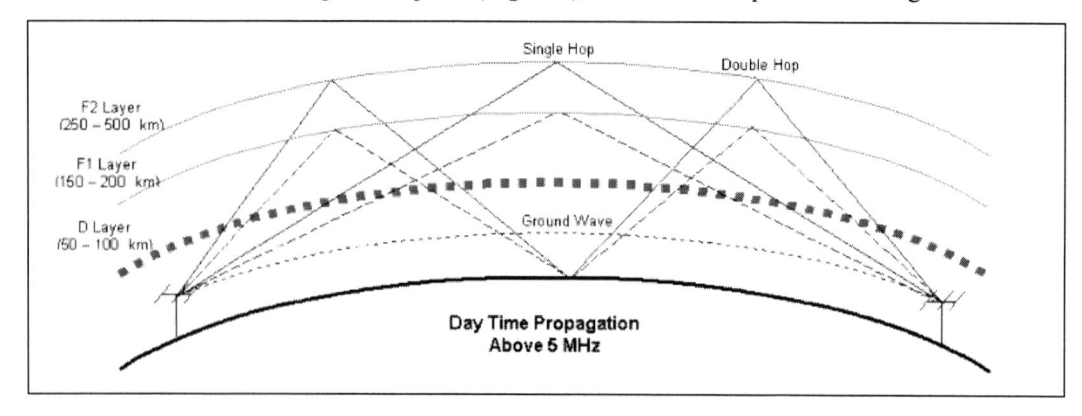

CHAPTER 4: DIGITAL MODES AND THE IONOSPHERE

Fig 4.14: Long Path DX propagation mechanism

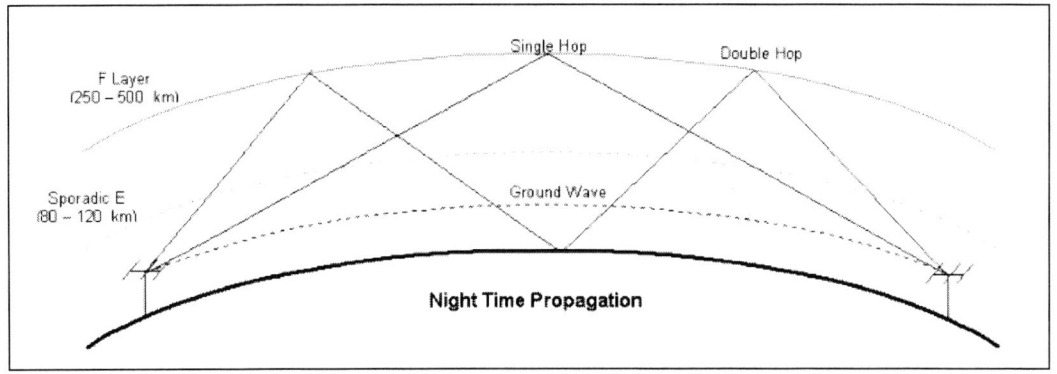

by the sun, the F-layer recombines and becomes more stable, while at the same time slowly losing its reflective properties. Signals propagated via the night-time path (which we call the 'long path') are often very strong and very stable, and are characterised by far fewer alternative paths than the day-time path. Because the path is generally longer than the day-time 'short path', the delay will be longer.

Not infrequently, both short and long paths are viable at the same time, leading to further possibilities for multiple delay reception. In this case the path differences can be very large. For example, on the path from Europe to the South Pacific the difference is about 50–60ms. Fig **4.15** illustrates both short and long path reception from Cyprus in the Mediterranean to New Zealand.

Figure 4.15 is a single frequency 'chirpogram' [7], which displays local time of day (horizontally) and signal time of arrival in milliseconds (vertically). The darkness of the image represents signal strength. Between 0500 and 0830 a strong signal is received with zero delay (relative to the shortest possible short path). Note that the signal also arrives during this period with various delays up to 20ms, representing other day-time path mechanisms (as previously outlined), and that these are generally weaker than the shortest path. The same applies between 1230 and 1800, although note that the shortest and strongest path was not viable until 1330.

Now look at the signal from 2200 to 0000, and from 1800 on. The recording

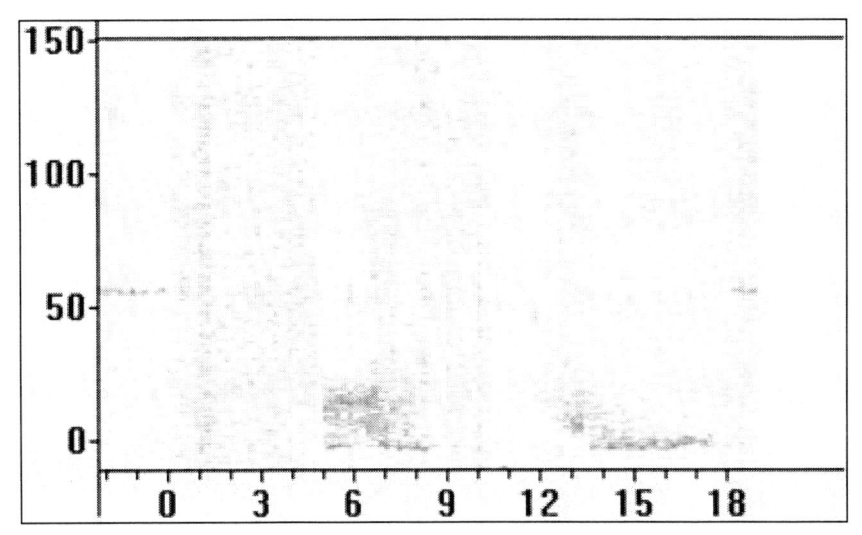

Fig 4.15: Long and short path chirp sounding

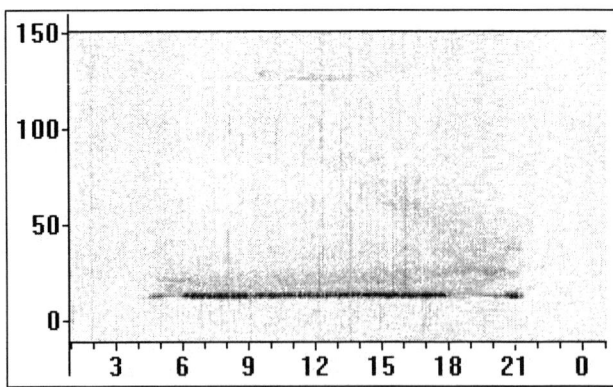

Fig 4.16: Long and short paths concurrently viable (Picture: Peter Matinez)

was not continued beyond 1900, but in fact this path was viable throughout the period 1800–0000. This is the night-time long path. Note the added delay, some 55ms later than the shortest path signal. The signal is strong and consists of one single stable line with a steady delay.

In **Fig 4.16**, recorded by G3PLX in the UK on 29MHz, the same sounder in Cyprus shows the short path at 12ms delay, with some scatter, and traces of the long path at 127ms, illustrating that long path and short path mechanisms can exist at the same time. Note also the scatter at around 1500–2100. Signals are known to pass right around the world several times!

Digital modes such as MFSK16 and MT63 have been designed with very slow baud rates, in order to combat multi-path delay effects. The MFSK16 symbol is 64ms long, and the first and last 5ms of each symbol are ignored, so path differences as much as 10ms can be accommodated without error. This mode also includes a symbol clock AFC system which can track slowly varying delays. MT63 uses symbols 100ms long (10 baud), and very strong FEC. Other advanced systems [8] use very high baud rates, but include a ranging signal embedded in the transmission which allows timing variations to be precisely measured and tracked.

Doppler Modulation Effects

Multiple ionospheric reception paths on HF have been described. As the day progresses, the ionisation present in the ionosphere slowly changes, affecting the refractive index and thus the effective height of the refractive layers. This effect is particularly strong around sunrise and sunset. In addition, there is agitation of the charged particles of the ionosphere which causes motion of a more localized but random nature.

This agitation can be thermal, weather induced (high-level winds) or caused by solar wind (solar particles). Sudden unusual changes in refractive properties are also caused by the sudden arrival of bursts of particles from the sun (solar flares).

Frequency Shift

Wherever there is motion, there will be Doppler shift. It matters little whether the signal source or the receiver is in motion, or whether the point of signal reflection is in motion, the effect is the same. The slow diurnal variations in apparent reflective height cause carrier frequencies to be shifted slightly (the effect is small), while the agitation effects cause small frequency- and phase-noise components to be added to the reflected signals. Solar flares cause quite spectacular frequency shifts of short duration.

The Doppler effects are frequency dependent, and so are not a problem on the lower HF frequencies. At 20m the slow frequency shifts can be several Hertz. **Fig 4.17** shows reception of a standard frequency transmission on 16MHz.

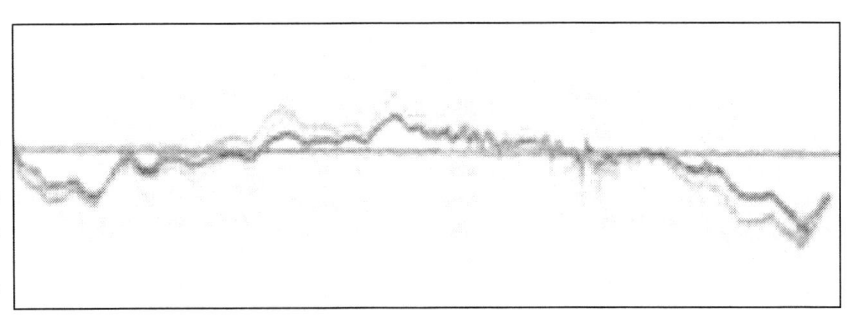

CHAPTER 4: DIGITAL MODES AND THE IONOSPHERE

The horizontal scale on this special high-resolution spectrogram is about 90 minutes, while the vertical scale is 10Hz span [9]. The comparatively straight line is a local 16MHz precision reference, and the wobbly line is the carrier of VNG (Australia) over a range of 3000km.

Fig 4.17: Slow Doppler shift on 16MHz

The carrier frequency of this standard-frequency station apparently varies by ±5Hz! In places there are two lines – these represent the 'ordinary' and 'extraordinary' rays, which have different propagation paths since they have different polarization [10]. Note also that just at the end of the recording the two rays coincide and then disappear, at the point where ionospheric propagation was lost. The other faint lines are 1Hz sidebands as a result of the seconds ticks on the VNG transmission.

Fig 4.18 shows WWVH reception on 20MHz over a range of 7000km. At this range WWVH (Hawaii) is not in the skip zone, so the signal is fuzzy because reception is by scatter, and subject to much more random Doppler modulation. The average frequency still varies in the same way as the previous illustration. In several places there are sudden peaks in the signal frequency, and one very big peak. These are caused by solar flares, which suddenly and dramatically lower the reflective layer for a matter of 10 minutes or more.

Phase Modulation

These frequency effects just described are interesting, but do not generally affect digital communications as they are small and relatively slow effects. However the companion Doppler-induced phase-modulation effects can be devastating. In general, PSK modes are transmitted using differential PSK, so the absolute phase is unimportant, and can vary slowly as the ionosphere changes. However the agitation of the ionosphere causes relatively fast phase-noise modulation of signals, especially in polar regions.

The effect is also frequency dependent, and affects modes with lower baud rates more than those with higher baud rates. This is because there is more time for the phase to change adversely if the symbol duration is longer.

For example, on 20m it is not uncommon for PSK31 signals received over the poles to be completely unreadable because the incidental noise modulation exceeds the intended phase modulation of the signal.

Fig 4.18: Doppler shift on 20 MHz

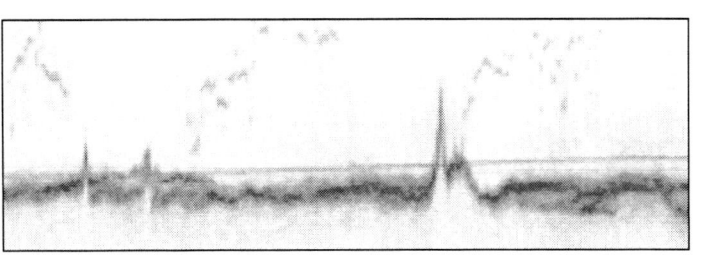

No amount of power increase will fix the problem. The only solutions are to use a lower frequency, a different path, a higher baud rate or a different mode.

MFSK16, PacTOR I and

41

RTTY are almost unaffected by Doppler phase modulation, since they are not phase coherent. PSK63F (not common, but just like PSK31 only faster) is also less affected since the baud rate is higher than PSK31. PSK-Hell, also a PSK mode, is almost completely unaffected by Doppler modulation, since not only is its symbol rate 245 baud, but the eye averages out the received phase errors.

Ultimately Doppler modulation on the higher HF bands limits the practicality of all narrow-band digital modes, even those that do not use PSK. On the lower bands this effect is not a problem, and on LF very narrow band modes can be used with no fear of Doppler shift or phase modulation.

Comparative Performance

It is very difficult to assess the performance of digital modes by direct comparison, as technologies vary as do ionospheric conditions. With the right equipment however, it is possible to make subjective comparisons between modes using an ionospheric simulator, which introduces noise, multi-path effects and Doppler effects in accurate and repeatable amounts (although not totally realistic). There is a very good website on the subject [11], from which the examples in **Table 4.3** have been extracted. The signals were all assessed with two simulated paths with 2ms differential path delay and 1Hz Doppler, at a signal-to-noise ratio of –10dB in a 3kHz bandwidth. This is a very difficult test for any mode!

Unfortunately this technique can only be used easily on simplex chat modes. ARQ modes are difficult to simulate due to the DSP processing delays and the need for a duplex path. To date no fuzzy mode simulations have been assessed.

Designing Modes for HF Performance

Now that we have reviewed the most important propagation challenges, we are in a position to assess what we would need to incorporate in a 'designer' digital mode. There are already quite a number of these specialized modes, and it

Table 4.3. Ionospheric simulation performance (courtesy Johan Forrer)

As transmitted
```
It should accept COM2, COM3, COM4 command line parameters
(default is COM1)
```

PSK31
```
It trould a oc?t Cr M2o  COOðõr MM cemmand line farao etera
```

PSK63F
```
ItjÑzld s faxt COM2, COM3Uc¶Re'e(nd line ®c'Wters (default is
COMq¥
```

MT63 1K
```
It should accept COM2, COM3, COM4 cofmand line arameters
(defa*f= {s[COM1)T
```

MFSK16
```
It should accept COM5, COM3, CIM4 command rine param ttsw
(default isto ei6
```

CHAPTER 4: DIGITAL MODES AND THE IONOSPHERE

Table 4.4. Designer Amateur modes

Mode	BW (Hz)	Speed (baud)	Designer	Purpose
FM-245	300	245	IZ8BLY & ZL1BPU	Weak-signal long-path DX QRM tolerant chat mode
FSKCW	2–10	0.05	VK2ZTO	Weak-signal LF beacon mode for FFT reception (using ARGO)
FSK441	2400	441	K1JT	Specialized high-speed meteor-scatter mode
MFSK16	316	15.625	ZL1BPU	Weak-signal long-path DX and NVIS chat mode
MT63	1000	640	SP9VRC	Weak-signal interference-tolerant broadcast mode
PSK31	50	31.25	G3PLX	Weak-signal DX chat mode with minimum bandwidth
PSK63F	100	62.5	IZ8BLY	DX chat mode with improved Doppler performance and FEC
Q15X25	2400	83.33	SP9VRC	Fast CSMA packet modem with high QRM tolerance

will be immediately obvious that the requirements conflict to some extent.

For example, to improve sensitivity and reduce impulse-noise damage, we need to use a narrow bandwidth and a slow baud rate. To control the effects of multi-path timing errors, we also need a slow baud rate. However, Doppler becomes a serious problem if the bandwidth is narrow or the baud rate is low!

In order to maximise data rate and minimise carrier interference, it is a good idea to use frequency diversity, ie spread the signal right across the available receiver bandwidth, and to use a very high baud rate. It also helps to use time diversity, by spreading the signal over time using an interleaver. However, by doing this, the sensitivity may be reduced and reception may suffer badly from multi-path reception and selective fading.

Table 4.4 illustrates some of the recent Amateur designer modes and the purpose for which they were intended.

References

[1] The highest part of the atmosphere, where reflective charged particles are generated by the sun.

[2] The lower part of the atmosphere, up to about 30km, specifically where the weather is.

[3] An upper area of the atmosphere where the air is thin, but meteorites leave reflective trails.

[4] That is, how much it responds to the peak of the signal relative to the average level.

[5] Q15X25, MSK441, STANAG 4285 for example.

[6] MT63 and MFSK16 for example.

[7] A passive ionogram received using a technique pioneered by Peter Martinex G3PLX.

[8] For example STANAG 4285.

[9] Fig. 4.16 and 4.17 should not be compared with other spectrograms in this book.

[10] It is as though the refractive index of the ionosphere is different for different polarizations.

[11] http://www.peak.org/~forrerj

5

History

In this chapter

- Text transmission
- Automatic reception
- Image transmission
- Amateurs involved
- Amateur designs
- Error-coding systems

ALTHOUGH AMATEURS have used digital modes for communications for over 50 years, and Morse for something like 80 years, the history of digital modes starts much further back. For example, what were beacon fires and the ringing of church bells in times of emergency, if not digital modes? Veteran RTTY experts G8GOJ and G8EXV maintain that the drums of Africa and smoke signals of prehistory should also be considered digital modes [1].

Seriously, however, the beginnings of digital modes came with the development of conducted electricity and electromagnetism in the early 19th century. Michael Faraday explained electromagnetic induction in 1831, and very soon there were systems with needles and coils of wire to deflect them, used to send primitive messages over reasonable distances using wires. James Clarke Maxwell published equations in 1865 to predict the existence of radio waves that travelled at the speed of light. These waves were first demonstrated at about 66GHz by Heinrich Hertz in 1888. J.C. Bose demonstrated EM waves in 1896, achieving a range of about 1.4km (not bad for 60GHz), using the signals to operate remote devices [2].

He also developed a wide range of quasi-optical microwave devices, and developed the first solid-state microwave detectors. Marconi achieved a range of 2km in 1894, transmitting the Morse letter 'S', and achieved trans-Atlantic transmission in 1901.

Fig 5.1: The Morse Sounder (Photo: Sam Hallas)

Fig 5.2: The Wheatstone ABC Telegraph in the British Telecom Museum (Photo: Sam Hallas)

Fig 5.3: The famous Samuel Morse Embosser in the Science Museum (Photo: Sam Hallas)

Text Transmission

William Cooke and Charles Wheatstone developed a telegraph system for sending letters and numbers based on five wires and five needles in 1837, but the cost of five separate wires for transmission soon led to improved two-needle systems with only two wires.

Samuel Morse's system also dates from 1837, but used only one wire and one needle, with the current changing direction to deflect the needle in different directions for key down and key up. This was an early double-current system, and it is interesting that Morse is more remembered for his code (very probably developed by his assistant Alfred Vaile), than for his effective signalling system.

In order to extend the useful range, the weak signals were amplified by sensitive electromagnetic relays. The clicking of these relays and the clicking of paper-embossing receivers soon led to the discovery that the signals could be read by ear, and the Morse Sounder (**Fig 5.1**) was born. The Sounder is simply a relay mounted on a wooden base to amplify the sound.

In 1840 Wheatstone introduced the ABC system, which used a single wire and a single receiving pointer, which stepped around as though driven by a stepper motor (**Fig 5.2**). The advantage of this system was that no skill was required – the operator simply pressed keys with letters and numbers engraved on them, or read the letter next to the pointer and wrote it down.

While the simplicity and reliability of the Morse code system eventually won the day, the importance of operational simplicity in the reduction of operator-training requirements was to surface again and again throughout the history of digital communications.

Automatic Reception

It soon became obvious that printed reception had many advantages. It made resending and transcription of messages easier, with better accuracy, and made faster transmissions practical. Several paper embossers and paper-marking machines were developed by attaching a stylus or pen to a relay armature. Perhaps the most famous was that of Samuel Morse (**Fig 5.3**), used to send a message from Washington to Baltimore in 1844 'What hath God wrought?' In 1845 Thomas John developed the definitive Morse Inker, which made marks like a square wave with a pen on paper tape moved along by clockwork.

Because of the cost of wire circuits (and later, radio equipment) it was important to use the circuits as efficiently as possible. For this reason transmissions were sent as fast as possible, and for this automatic printing was important, allowing the messages to be printed at speed and read later at leisure. Later tape-driven automatic transmitters further enhanced transmission speeds. The Wheatstone Morse tape perforator dates from 1867, and allowed several operators to cut tape 'offline' for high-speed transmission on a single circuit. Tapes could also be edited and corrected with a pair of scissors and a pot of glue.

Fig 5.4: The Baudot Distributor – British Telecom Museum (Photo: Sam Hallas)

Making maximum use of each circuit also led to the development of multiplexers, which allowed several operators to key directly on one circuit, by allocating each a time slot in turn. Emile Baudot developed a highly-reliable rotary distributor (**Fig 5.4**) for a multiplexed system in 1874. This device was a motor-driven switch that maintained accurate synchronism at the other end by passing small DC currents across the circuit. In addition to multiplexing four or more operators on a single wire, this distributor multiplexed five keys for each operator's fingers on the same single wire!

The five key system is significant, because in all on/off combinations of five keys, there are 32 possibilities. With a convention to define a set of letter codes and another set of number and punctuation codes, and a convention for switching code sets, it is just possible to define enough characters for effective message transmission with a five-unit code [3].

Fig 5.5: The Baudot Keyboard in the British Telecom Museum (Photo: Sam Hallas)

The Baudot Multiplex operators used a five-key piano-like keyboard, hand keyed in time with a clicking sound from the distributor (**Fig 5.5**). Keys were pressed one or more at a time at a very precise rate in order to keep in 'cadence' with the distributor. The keys remained locked down until the end of the operator's time slot. The receiver stored incoming pulses from the receiving distributor on electromagnets, and directly-printed letters and numbers on a paper strip. The system was used on the cable between London and Paris from 1897.

Since the text was manually transmitted by piano keyboard, the code was designed to be easily memorised, with the least use of the fingers for the most common characters. No control characters were used other than the letters and figures shift, as the printout was on paper strip, and no page control was necessary. This was the first major use of a five-bit code with a shift technique.

This code may have evolved into ITA1, which was used on a very few early text printers. **Fig 5.6** shows the Baudot keyboard and alphabet. Keys 1, 2 and 3 were pressed by the first three fingers of the right hand, and keys 4 and 5 by the first two fingers of the left hand. Each black dot represents a key press, at which time current flows (MARK).

Note how the figure shows that the code was designed in the manner of a 'grey code', where just one bit differs from one letter to the next.

Many five-hole paper-tape punches and readers were developed for the Baudot and later systems. These permitted fast direct printing using a distributor system and a Baudot printer. However the alphabet Baudot developed bore no resemblance to ITA2, and was not used for radio transmission [4].

There were other methods used to send and directly print text. For example, David Hughes developed the first synchronous direct-printing telegraph in 1845. This system used what we would now call pulse-position modulation, where the delay between pulses defined which character was to be printed.

Donald Murray, a New Zealand sheep farmer turned journalist and inventor, worked in Paris, developing techniques for automatic typesetting. He devised his own tape reader, and a typewriter-like keyboard perforator. The transmitter used tape-reading contact levers which operated a distributor in a similar manner to the Baudot multiplexer. His system was highly successful, and he won several major prizes in 1902 and later. He sold the US patent rights to Western Electric.

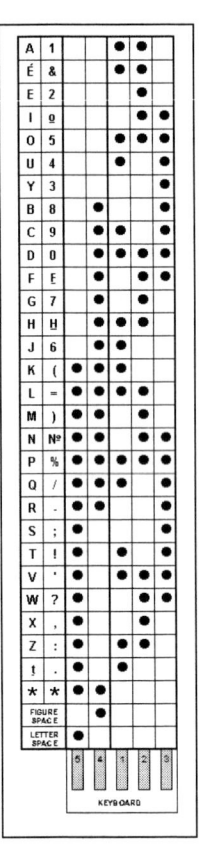

Fig 5.6: The Baudot keyboard and alphabet

Murray developed a new alphabet for his five-bit system, which minimised wear and tear by using codes that involved fewer mechanical movements for most used characters. Murray's code corresponds in all details of letter and number coding to the ITA2 alphabet. However, because Murray's system was synchronous (like Baudot's), the start and stop bits now associated with ITA2 were not used.

An interesting aspect of the Murray code, which evolved as ITA2, is that several of the control characters are 'symmetrical'. If for some reason the paper tape is reversed or the connections to reader and punch reversed, while the text that prints will be gibberish, the machines will appear to behave normally. The LTRS and FIGS characters are reversible, while the CR becomes LF, and vice versa. Quite how this evolved and what use it might have been is not known.

The first really successful start–stop (asynchronous) system was patented by Howard Krum in 1910. The Krum machine directly transmitted from the keyboard, and its success led to the formation of the Teletype Corporation. The now familiar start–stop RTTY system was well established by 1920 and the ITA2 standard for asynchronous transmissions was ratified by the CCITT in 1933.

The start–stop technique was a very important development, as it allowed the precision of machine timing to be relaxed considerably. As each character started, the machines resynchronised, so a motor-speed error of 0.5% would not cause printing errors, and a simple governed motor would suffice to control each machine. Synchronous machines had required very high operating precision, and some means of keeping the machines in step, in order to maintain synchronism accurately throughout the transmission.

Frederick Creed sold his first batch of keyboard-operated perforators to the British Post Office in 1902. By 1922 Creed machines incorporated the start–stop system and ITA2 alphabet, and the Model 3, combining keyboard transmitter and gummed-tape printer, became the standard machine for the new Post Office telegram service (**Fig 5.7**). The famous Model 7 was the first to operate at the standard 50 baud, and was introduced in 1931. Many are still in working order.

Fig 5.7: The 1927 Creed Model 3 Telegram unit in the Science Museum (Photo: Sam Hallas)

Image Transmission

The history of electric image transmission is at least as old as that of text transmission. Alexander Bain patented a pendulum-operated image transmission system in 1843. This used chemically impregnated paper that changed colour as current passed through it. In addition to this first facsimile machine, he also developed a chemical Morse printer in 1845. Bain was an apprentice clockmaker, and his machines used manual synchronism of the pendulums at the start of transmission. Transmission was limited to text and simple shapes, using metallic type brushed by a pendulum-driven stylus. The pendulums were kept swinging accurately by timed electrical pulses. His system was first used commercially on a circuit between Paris and Lyons in 1865.

Improvements like the rotating paper drum and lead screw developed by Frederick Bakewell in 1850, and the photoelectric reader patented by Dr Arthur Korn in 1902, led to the familiar FAX machine used first for press pictures in 1906, and transatlantic transmissions in 1928. This technology was in use for press photos up to the 1980s, and is still used today for weather pictures. Radio FAX transmissions predate radio teletype (RTTY) by about 15 years.

In the 1920s, start–stop teleprinting machines were large, noisy, unreliable and very expensive. Dr Rudolf Hell worked in the printing industry in Berlin, and saw a need for small, lightweight and inexpensive machines for use in newspaper offices to allow copy to be distributed from a central agency. He developed a paper-tape printing system which met these requirements, and used many innovations and technologies far in advance of other equipment of the time (**Fig 5.8**).

Rather than sending a train of pulses representing a character code for each letter, the 'Hellschreiber' used a system of pulses that directly described each character, in a

Fig 5.8: The keyboard of Hell's text transmitter

Fig 5.9. Transmission from a 1940s Hellschreiber

scanned manner, and a printing system where the transmitted pulses operated a scanned inking system on paper tape (**Fig 5.9**). The press receiver received at 40 WPM, required no synchronism, and contained only three or four moving parts! The system was first patented in 1927, and was combined successfully with radio in 1937, when it was used in the Spanish Civil War by the German Condor Legion. Reuters and other press agencies world wide used the 'Hellschreiber' system for nearly 50 years. It was eventually superseded by the Telex service as the cost of start–stop teleprinters was reduced.

The US Army RC-58B system of 1944 was similar to the Hell system, but used optical scanning to read hand-sent text on a paper strip, and could therefore easily retransmit received messages. The receiver was almost identical to the Hellschreiber. FSK transmission was used.

Slow-Scan Television is a truly Amateur invention, developed by Copthorne McDonald W0ORX and others in the early 1950s. Early transmissions were made using AM transmitters and static images generated by flying spot scanners. The pictures were received on cathode-ray tubes, and so were very like fast-scan TV slowed down. The present FM transmission standard was proposed by McDonald in 1961. Fully electronic frame-store systems, colour transmissions and live camera pictures were introduced in the 1970s and 1980s. SSTV remained an expensive fringe activity until the introduction of computer-based systems in the early 1990s. The present single-frame picture system became popular at about the same time.

Amateurs Involved

RTTY was the first of the conventional digital modes to be used by Amateurs. Although developed around the turn of the 20th century, RTTY was not successfully used by radio until the 1940s, with the development of FSK. RTTY was first mentioned in the ARRL Handbook around 1955. Interest increased as war-surplus and the older class of mechanical teleprinters became available through the 1960s, with a peak in the 1970s coinciding with the widespread availability of commercial SSB transmitters, and another peak based on the first MCUs in the early 1980s.

Many an operator started with a Creed Model 7 or Teletype Model 15 and a home-brew Terminal Unit (**Fig 5.10**). Early Amateur teletype was therefore characterised by large, heavy and noisy mechanical machines, and relatively large and complex decoding equipment. There was no practical means of recording messages electronically, so fixed messages such as CQ messages and the inevitable 'brag tape' were recorded on punched paper tape, and later on audio cassette.

Few amateurs migrated to modern commercial teletype equipment, most preferring to go directly to the 'glass teletype', or computer-based generation and reception of RTTY, using an adapted RTTY modem to convert the signals to and from the radio and the computer.

Hellschreiber did not move into the Amateur world until the 1970s, when some war-surplus machines surfaced, and the German Post Office released further machines. Hell operation was therefore always hampered by lack of equipment until the development of fully electronic and computer-based solutions in the 1980s. Facsimile has always been, and remains, a fringe activity among Amateurs. Perhaps the most widespread use in Amateur circles is for reception of weather-satellite images on VHF.

Most other commercial digital systems have been completely ignored by Amateurs. The only other commercial system to be used seriously by Amateurs was SiTOR, adapted by Peter Martinez G3PLX as AmTOR in 1978, and used world wide until about the mid 1990s. SiTOR was developed for the maritime service in the early 1960s, and is still in use today.

Fig 5.10: The popular Creed Model 7 Teleprinter in the British Telecom Museum (Photo: Sam Hallas)

Amateur Designs

Arguably the first Amateur 'designer mode', PSK31 was developed by Peter Martinez G3PLX in the mid 1990s as a replacement for RTTY. PSK31 was intended to provide better performance with lower power and narrower bandwidth than RTTY, and yet be as fast and easy to use. It became widely available in late 1999 with the introduction of a sound-card software version, and use of PSK31 now far outstrips all other digital modes.

PacTOR, developed by Hans-Peter Helfert DL6MAA and Ulrich Strate DF4KV, and Clover, developed by Ray Petit W7GHM, are examples of Amateur-developed modes that have since became commercial systems.

While Hellschreiber no longer has any commercial application, it has much to recommend it in Amateur service. The mode has been enhanced significantly by the introduction (by Peter Martinez G3PLX, Lionel Sear G3PPT and Nino Porcino IZ8BLY) of DSP grey-scale processing in the late 1990s, and by the more recent development of highly sensitive and robust PSK-Hell versions by the author and Nino Porcino. Few modes are simpler to use, or offer the performance and pleasure of this simple mode.

The latest developments have taken the concept of designer modes seriously, addressing the specific needs of individual applications. For example, MT63, designed by Pawel Jalocha SP9VRC and ported to Windows by Nino Porcino IZ8BLY, provides high-speed transmission in heavy QRM; MFSK16 by the author and IZ8BLY was designed for low-power long-path DX, and also performs well on noisy low-HF bands at night; FSK441 was designed by Joe Taylor K1JT for meteor-scatter propagation; Q15X25 was developed by Pawel Jalocha SP9VRC, Timo Manninen OH2BNS and others as a modern robust replacement for AX25 and TCP/IP [5] packet on HF. Digital speech, improved

IP and AX25 modes and even fully-digital high-definition SSTV are under development. At the same time old and new modes are also migrating to newer computers and operating systems.

Obviously, new modes are proposed and developed to overcome propagation limitations or to provide improved performance. One important consideration when developing new designer modes is ease of use. In this regard, it is important that each new mode be uniquely identifiable; for example it should sound different, or look different, when viewed on a waterfall or other tuning display. While modes that look or sound similar to existing modes may have advantages, without a means of differentiating them they are useless except for experimentation.

Error-coding Systems

In 1948 Claude Shannon showed that controlled redundancy added to digital communications could potentially reduce errors in reception at low signal-to-noise ratios. Error-control coding (ECC) techniques use controlled redundancy to detect and correct errors [6]. Many famous names in the world of mathematics and error coding have contributed to the body of ECC technology used by Amateurs today. Richard Hamming [7], father of error-correcting codes, Andrew Viterbi [8], developer of a highly-efficient convolutional decoder, J.B. Moore [9], John Meggitt [10], developer of an efficient systematic decoder, Phil Karn KA9Q [11] and many others have directly or indirectly contributed to the ECC digital modes we use today.

References

[1} *A Short History of Telegraphy* by Alan G Hobbs, G8GOJ, and Sam Hallas, G8EX. http://www.samhallas.co.uk/telhist1/telehist.htm.
[2] http://www.tuc.nrao.edu/~demerson/bose/bose.html.
[3] http://www.nadcomm.com/fiveunit/fiveunits.htm for an excellent treatise on five-unit codes.
[4] Explained in Chapter 2.
[5] TCP/IP – Terminal Control Protocol / Internet Protocol.
[6] For an excellent introduction to error coding, see the web site http://www.ittc.ukans.edu/~paden/reference/guides/ECC/introduction.html.
[7] Bell System Technical Journal 'Error Detecting and Error Correcting Codes'. See http://www.engelschall.com/u/sb/hamming/
[8] Co-founder and director of Qualcomm.
[9] Proposed a special alphabet and use of ARQ for HF use (SiTOR).
[10] Cambridge graduate, worked for IBM.
[11] Developed ECC systems for space probes and algorithms for Amateur ECC use. Works for Qualcomm.

6

Getting started

In this chapter
- What you need
- Setting up
- Choosing software
- Special operating considerations

THERE'S NO NEED to be daunted by the prospect of operating digital modes. Certainly there are a number of very complex systems that take some care and experience, but there are also some very effective modes that are easy to set up, and a pleasure to use.

What You Need

You need some sort of interface for the radio signals, you need a radio transceiver, and a computer (**Fig 6.1**). First we'll review the different types of interface systems for digital modes:

- TU and simple PC
- MCU with terminal or simple PC
- Advanced MCU with special PC software
- Comparator interface with special PC software
- Sound card, fast PC and software

The last option is the best. This is because the PC sound-card interface is by far the simplest way to get started, and is the cheapest approach since most families now have a computer, and there's nothing more to buy. There is a huge range of excellent sound-card radio software, most of it free.

For all of these systems you need an HF SSB transceiver, and access to the receiver audio output and microphone audio input. It is best if this is a modern solid-state unit, with good filters and low drift, but many operators use older rigs for RTTY, MT63, Hellschreiber and SSTV with no particular problems. These are the modes less affected by drift and poor frequency netting.

Some of the newer modes require very high stability, and also very low fre-

DIGITAL MODES FOR ALL OCCASIONS

Fig 6.1: A typical digital-modes ham shack (the author in residence)

Fig 6.2. The Maplin TU1000

quency offset between transmit and receive. Most synthesised transceivers will suffice. If your transceiver drifts less than 5Hz per over (they usually drift back on receive) and has less than 5Hz offset, you should be able to use it very successfully for the newer modes. Unfortunately offset is quite difficult to measure, and cannot be accurately corrected by using the Receiver Incremental Tuning (RIT). Those you talk to on the air will soon tell you if the offset is excessive!

You also need a computer, and a means of connecting it to the transceiver. With some older systems a terminal or simple PC will suffice, but you will be missing out on much of the fun with that approach, as almost all the new applications have been for fast PC and sound card. Details of how to use the PC and sound card will follow later in the chapter. First, for completeness, we'll review the other options, although with a few exceptions they are no longer recommended.

TU and Simple PC

The Terminal Unit (TU) is a modem (MOdulator/DEModulator), with no microprocessor or other intelligence. It simply converts the digital signals from the computer into sounds, and those from the receiver back into digital signals. The TU relies on special software in the PC to interpret what is received, generate the data to be transmitted, and control operation.

There were some famous TU designs, such as the ST-6 by Irv Hoff (1960s), the BARTG Versaterm and the Maplin TU-1000 (**Fig 6.2**) kitset, both from the 1990s. These could be easily adapted to operate AmTOR and even PacTOR, as well as RTTY.

There were many computer programs for TU-based RTTY. One of the best was BMK-Multy by Mike Kerry G4BMK, which even operated AmTOR and PacTOR. The author used an adapted TU-1000 with BMK-Multy successfully for many years. However these applications have been surpassed now by software using modern computers, graphical displays, sound cards and digital signal processing.

Fig 6.3: The Kantronics UTU

MCU with Terminal or Simple PC

The first Multi-mode Control Units (MCU) were (by present-day standards) very simple. They consisted of a conventional modem with an additional simple microprocessor, used to convert ITA2 to ASCII and vice versa. They could communicate with a 'dumb terminal' in ASCII, or with a PC using a simple 'terminal program'. The Kantronics UTU (**Fig 6.3**) is a good example of an early MCU which operated RTTY, AmTOR and Morse. Another early example was the unit devised by G3PLX for AmTOR.

Later units had better performance and other modes, such as AX25 packet for VHF and HF. The most widely known of these was probably the AEA PK-232, which was available in many versions for over ten years. These units could be used with a 'dumb terminal' but very often came with matching PC control software, allowing modes such as FAX and HF/VHF bulletin boards to be used. Once again, these units have been passed by in the rapid advance of PC DSP software.

Advanced MCU

The commercial modes Clover, PacTOR II (**Fig 6.4**) and G-TOR are only available with specialised modems. These units contain microprocessors designed for fast digital-signal processing, so all the filters and modulation processes are carried out mathematically. These modes are very complex and offer very high performance. In most cases the program used by the special processor is 'downloaded' to the unit by the special software that runs on a PC in conjunction with the MCU. This means that the MCU can be completely configured in an optimum way for each mode, and the MCU can be updated easily when new versions or new modes appear.

Unfortunately these advanced MCUs are very expensive, and although widely used commercially, except for high-performance bulletin-board networks, few Amateurs can afford to use them. In general they only operate one advanced proprietary mode – for example you will not find Clover 2000 and PacTOR II on the same MCU. Some do however offer freely-available modern modes for added convenience (for example PSK31).

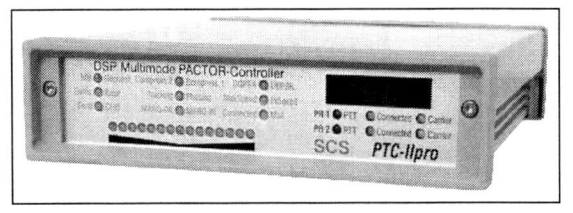

Fig 6.4. The SCS PTC-II (Photo: SCS)

Comparator Interface

This technique was popular throughout the 1990s, and is still one of the cheapest ways to get started. The receiving interface consists of nothing more that a simple operational amplifier wired as a comparator, while the transmit circuit is just a low-pass filter and one transistor to operate the PTT. The whole interface usually operates from, and is usually powered by, a PC serial port [1]. Perhaps the best known interface of this type is the Hamcomm interface, although there were many others, including one by SSC which could be used for weather satellite reception. Software was also available for other older (non IBM PC) computers.

The disadvantage of the comparator type is that it generally only works on FSK signals, and performs poorly when signals are noisy. All the work is done in the PC, and there was a remarkable range of DOS-based comparator interface software for the PC-RTTY, SSTV, FAX, even Hellschreiber and spectrum analyser programs. Very few Amateurs use this interface seriously now, although it is still used by listeners for 'browsing', and for portable use with older laptop computers. Because of timing problems, there is no Windows™ software for the comparator interface.

PC and Sound Card

This is much the best and simplest way to get started. The PC uses its sound card to transmit and receive the signals. This approach requires the least equipment, provides the best performance, the widest range of applications, and (assuming you already have the necessary computer) is the lowest cost, as most of the software is free.

Check your computer against the table below. Virtually any home computer purchased new in the last five years will suffice. Some of the more advanced programs may require a 200MHz or faster computer, especially if operating in conjunction with a logging or bulletin-board program.

The newer Windows operating systems (2000, ME and XP) can sometimes be used if care is taken to use the correct sound-card drivers. There is a growing library of software for the Linux operating system, including PSK31, MFSK16, Q15X25 and STANAG 4285.

A limited amount of sound-card software is available for DOS, although fewer configurations are reliable, principally because of the difficulties experienced with different sound cards.

PC and Sound Card Computer requirements

Processor	Pentium 100 or better
Memory	16MB, preferably more
Display	VGA, preferably at least 800 x 600 SVGA colour
Hard disc	Not important, but perhaps 20MB required for all the programs you will need
Sound card	Soundblaster™ 16-bit type or similar
Serial ports	One required for rig control if VOX is not used
Operating system	Windows95™ or Windows98™ preferred

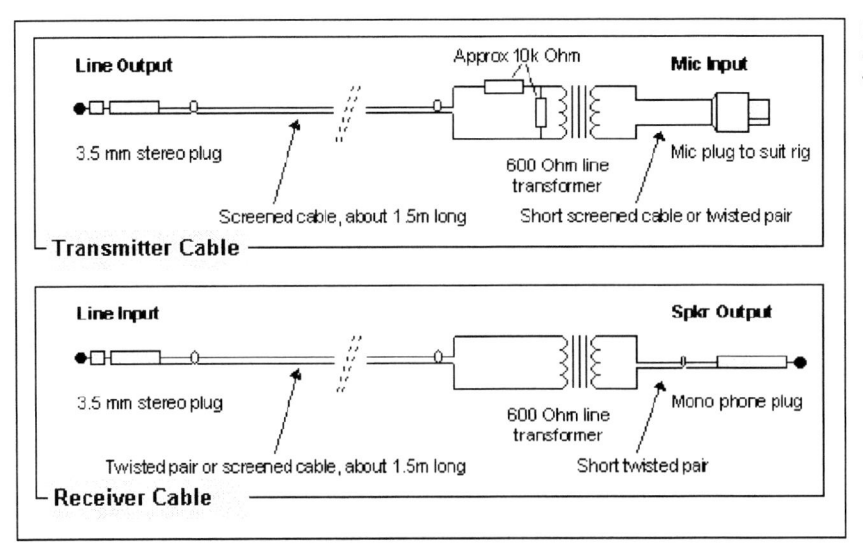

Fig 6.5: The simple cables used with the PC sound card

Almost no software is available for Macintosh computers, and even less for other types. The software used with older home computers such as the BBC and Acorn compares very badly in features and performance with modern PC software using DSP techniques and a sound card.

Connecting Up

With the computer and sound-card approach, literally the only other thing required is a method of connecting the computer sound card to the transceiver. The cables are very simple, and most Amateurs should be able to build their own, with shielded cable and a few simple components.

See **Fig 6.5**. The resistors in the transmit cable are used to attenuate the sound-card signal so that it does not overload the transceiver. If you use the microphone socket, a lower value of resistor may be necessary across the transformer. If an accessory socket or 'phone patch' socket is used, the values shown may suffice.

While the receiver cable is shown with a plug to be directly plugged into the external-speaker socket on the receiver, this isn't a very good idea if the speaker is disconnected in the process. It is best in this case to use an adaptor allowing both the PC and an external speaker to be connected, or to modify an external speaker by adding a socket for the cable.

Using PC Speakers

If plugging in the receiver cable disconnects the receiver loudspeaker, why not listen to the receiver via the PC sound card and speakers? Well, the main reason is that by connecting (in the mixer software) the Line In signal from the radio to the Line Out or Speaker Out and the speakers, you will also be sending the signal to the microphone input of the transceiver, and all sorts of problems will occur, especially if you plan to use VOX. Another reason is that very often the computer has only one output (Line or SPKR), so you can't use the computer speakers at the same time as the transmitter cable. This is most often the case with laptop computers.

DIGITAL MODES FOR ALL OCCASIONS

The Importance of Isolation

Why do the drawings in Fig 6.5 show transformers? Well, these provide complete DC isolation between the computer and the radio transceiver. This is a good idea in order to keep computer noises and hum out of the radio, but the most compelling reason is to prevent serious damage to the radio and the computer if the power-supply cable to the transmitter becomes loose, and the full 20A transmitter current passes through the microphone circuit, down the cable and through the computer sound card to ground via the PC power cable. Even if the power cable is considered reliable, significant current could still flow through the sound-card cable, causing instability, hum and RF feedback.

The transformers used can be line transformers (say from old telephones or modems), 600 to 1000 ohms, or transistor radio inter-stage transformers, which are usually 3000 ohms. These transformers are readily available from hobby electronics stores. The transformer impedance is not critical, and you should simply ignore any centre-taps on the windings. Check that the transformers are wired correctly by checking for no DC continuity from one end to the other.

It is helpful to check that the DC power supply for the transceiver has electrical isolation between the AC ground and the DC output. The supply and its case must be connected to the AC ground for safety reasons, and the DC supply negative line will be grounded to the radio earth via the transceiver case. There should be no connection between them in the power supply. These two ground systems should remain separate for many practical reasons [2].

There are many interface designs around that are not isolated, or only partially isolated. They are perhaps OK for receiving, but for transmitting they are not to be recommended – use them at your own risk!

VOX and PTT Control

Most operators, including the author, find VOX operation of digital modes quite appropriate and reliable, although the delay may need to be set longer than for Morse or SSB. If for some reason direct control of the rig is necessary, the transmit control must also use an isolated circuit. An opto-coupler does this nicely, driving the Press-to-Talk (PTT) directly without requiring a relay or any further power supplies. Direct PTT control must be used if the rig has no VOX, or if change over must complete before transmitter power is applied, for example when a power amplifier or transverter is to be used.

The digital-mode software usually controls the transceiver via a serial port, by driving RTS or DTR (often both) positive on transmit, with an appropriate delay before sending tones out from the sound card. The design in **Fig 6.6** is an appropriate PTT circuit for a transceiver with positive voltage on the PTT line and a current when PTT is closed of 10mA or less.

Fig 6.6: Isolated PTT control (10mA)

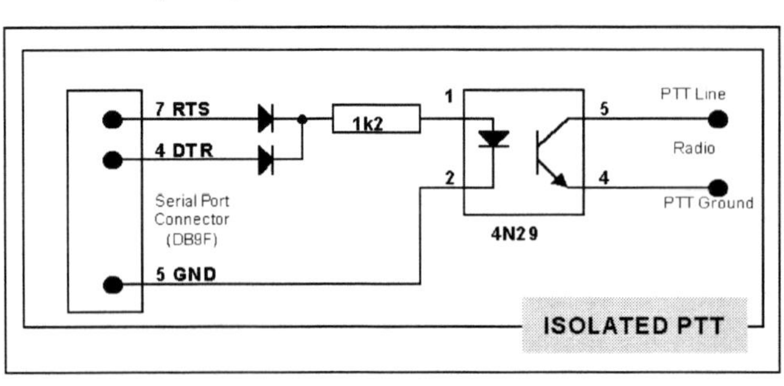

Many older transceivers have significant current flowing in the PTT circuit (perhaps in excess of 100mA). For transceivers such as these, use the opto-coupler to drive another NPN transistor, as shown in **Fig 6.7**. Many transistors would be suitable, for example the BC547B.

This circuit is also suitable for keying negative voltages up to 15V if the output terminals

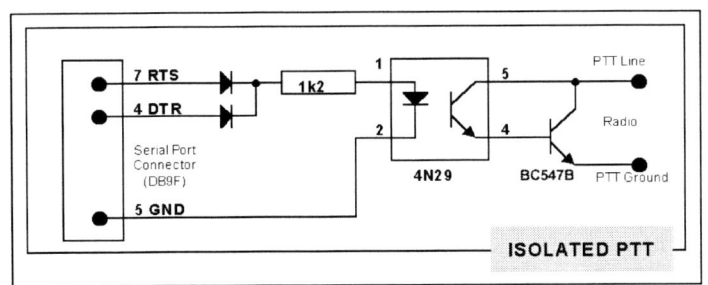

Fig 6.7: High current PTT control (>100mA)

PTT Line and PTT Ground are swapped. These circuits are not intended for switching inductive loads or voltages higher than 25V.

One of the problems with serial-port PTT control is that the state of the serial-port control lines RTS and DTR can be undefined until the correct software is operating. The transmitter may therefore operate unexpectedly. It is a good idea to include a switch to disable the interface while the computer is starting up. If you do elect to use VOX, be careful that the sounds the computer makes during start-up (and sometimes during operation) are disabled, so that you don't inadvertently transmit them! These 'system sounds' can also cause confusion by taking over control of the sound card, so they are best left disabled while operating on HF.

Using Phone Patch Connections

Many transceivers include 'phone patch' connections on the back panel, or perhaps on an 'accessory socket', offering line-level audio inputs and outputs for transmit and receive. Using these instead of the microphone socket and speaker socket can be really convenient, but there are a number of implications which have to be considered.

♦ Frequently the phone patch input will not operate the VOX

♦ There is often no PTT connection to the phone patch or accessory socket

♦ The Patch Out audio may or may not be controlled by the receiver audio control

♦ The Patch In audio may or may not be controlled by the microphone gain

♦ Very often the microphone is not disconnected when the patch is used

♦ The Patch audio is not isolated, so transformers will still be required

♦ Sometimes the Patch In audio is disabled unless the microphone is unplugged

♦ Leaving the Patch or Accessory connections in place for SSB operation may cause trouble (TX hangs up, unwanted noises on TX, VOX trips unexpectedly)

♦ It is not as easy to arrange an instant-switching microphone system (see below) using an accessory or Phone Patch connection

♦ If two identical plugs are used for the Phone Patch to computer interface, make sure they are labelled clearly!

Commercial Interfaces

Some Amateurs are not confident about their ability to make the cables and prefer to buy a suitable unit. Fortunately there are now several companies making

Fig 6.8: Sound card interfaces (Photo: West Mountain Radio)

sound-card interface equipment for HF transceivers (**Fig 6.8**) [3]. Some come with all the cables required and even the operating software.

Be sure to ask the supplier if their equipment is fully isolated. The unit supplied to the author has isolated transmit and PTT circuits, but did not provide the receive circuit at all. The manufacturer inappropriately suggested that a standard audio cable be used. Remember that all circuits between the transceiver and the computer must be isolated for safe operation.

Using a Microphone

Some commercial interfaces, such as pictured in Fig 6.8, and some home-made designs, have the additional advantage of a built-in microphone socket. Most digital-mode operation does not require the use of a microphone, so having to disconnect it to connect the interface is no big inconvenience. However, there are occasions when a design with a microphone is very convenient:

♦ Operating SSTV (chatting between pictures)
♦ Providing assistance to beginners, where you need to talk between overs to provide advice
♦ Testing new software or designs with a friend
♦ Operating a combined voice and data net (eg. using APRS, ALE or data modes for a communications exercise or emergency)

The interfaces with microphone are very simple to operate – the interface is connected in place of the microphone and the microphone is connected to the interface. To use them for digital transmissions, just ignore the microphone. To operate SSB, simply pick up the microphone and press the PTT to talk. It is usually best to turn the VOX off for SSB use, as the delay set for digital modes may be too long, and depending on the design, it may not trip correctly on voice.

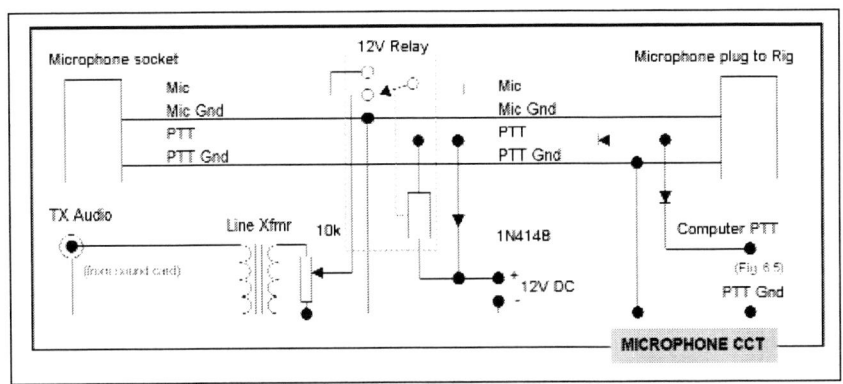

Fig 6.9: A simple microphone switching circuit

The interface between the computer and the transceiver was described in Fig 6.5, Fig 6.6 and Fig 6.7. The simplest way to add automatic microphone operation is to add a small relay and microphone socket to the PTT interface in Fig 6.6 of Fig 6.7. See **Fig 6.9**. This circuit and Fig 6.7 together replace the transmit cable in Fig 6.5, and also provide the necessary isolation. With a sensitive relay, the circuit will run for months from a small 9V battery. Current only flows when the microphone is active.

Build the circuit of Fig 6.9 and one of the isolating circuits of Fig 6.5 or 6.5A into the same small box. The relay connects the digital path when idle, and connects the microphone circuit when the microphone PTT button is pressed and the relay pulls in. This also actuates the transceiver PTT circuit. The computer PTT operates the transceiver, but not the relay, due to the action of the isolating diodes.

By the way, don't be tempted to operate digital modes by feeding the audio into the microphone itself via a speaker. While it might work, there's no likelihood that the frequency response will be flat or that the phase response or signal delay will be constant. Although some Amateurs will tell you otherwise, the results will almost certainly be disappointing.

Setting Up

Connecting up the PC based digital-mode station is really very simple. As shown in Fig 6.5 and **Fig 6.10**, connect the receive cable from the PC sound-card Line In to the transceiver audio output, bearing in mind the need to be able to still hear the audio from the receiver. Use the computer Mic In if there is no Line In.

Connect the transmit cable from PC sound-card Line Out to the transceiver's Mic In or accessory socket. Use SPKR Out from the PC if there is no Line Out. Connect the PTT cable (if used) between the PC serial port and the Mic In or accessory socket, where it shares a plug with the transmit audio cable.

The sound card can be easily identified on the computer back panel,

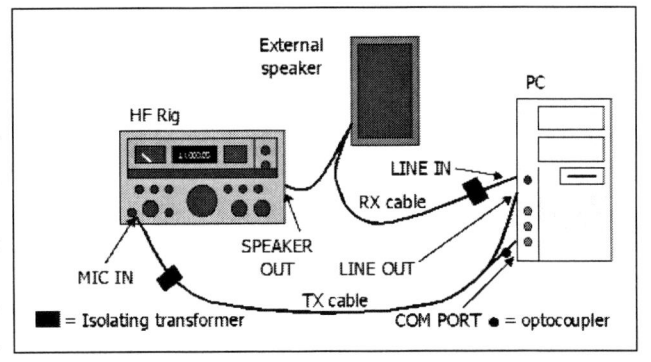

Fig 6.10: Connecting up the equipment

because it invariably has several mini-phone sockets and a 15-pin DIN type MIDI connector. This has a rounded trapezoidal shape, and 15 pin holes in two rows. (The video connector will probably have 15 pin holes, but in three rows). Some newer computers and most laptop computers have the sound features built in, so you will need to look closely to identify the connectors, or best of all, consult the user manual.

Adjusting the Sound Card

The adjustments are all performed in software, and generally once set for one mode or program the settings should be correct for all the rest. There are a very few multi-carrier modes where you might need to increase the microphone gain to achieve sufficient transmitter power (MT63 and MT-Hell for example).

There are two main software adjustments, for transmit and for receive, and it is not very obvious where to find these, especially the receiver adjustments. The better applications provide direct access to the adjustments. In addition to the gain settings, you need to select the correct inputs and outputs, and disable those not being used. The procedure and these adjustments are carefully and thoroughly described in Appendix D.

QRM in the Shack

Computers can be a serious source of interference in the ham shack. Apart from choosing a suitably quiet computer (and monitor, since they are often the main culprits) there are a number of things one can do to minimise interference.

First, the modern ATX style computer cases are better shielded than many earlier ones, and generally create less noise. Modern monitors are also generally better than older ones. Use a separate AC power feed for the monitor, the computer and its accessories, with a good line filter and spike catcher, and also a filter on the phone lead to the telephone modem, if used. Use a separate AC power feed for the HF transceiver, and its accessories, with another line filter. Keep the computer and the radio well apart, and keep the power wiring well away from the antenna feeders. It is a very good idea to bring the antenna-feed cables in one side of the shack, directly to the radio desk on that side, and keep the computer on a desk on the other side, near the AC power supply.

Make sure that the antenna tuner, all the antenna feeders and the cables between the radio and computer and other accessories, are well shielded [4]. Make sure that shields on RF cables and their connectors are very secure at both ends. Cables between computer and transceiver should be minimised (only the two or three isolated cables described above) and may benefit from clip-on ferrite filters.

A separate ground for the radio equipment is ideal, preferably with a series of stakes outside, and a short fat lead into the shack, to connect up the equipment. A large grounded metal sheet under the equipment can also help. A well-grounded steel desk is ideal. The DC supply for the rig must, for safety reasons, be earthed to the AC supply earth on the AC side, but the DC output must be isolated from the AC earth and should be connected directly to the radio ground. Other power supplies, such as those powering TNCs and MCUs, should also be isolated. Don't use the same power supply for radio accessories and computer accessories.

It may make operating less convenient to keep the equipment apart as suggested, but with a wheeled swivel chair you can quickly move back and forth between computer and radio. Above all, experiment with the setup for least noise. It will always be a compromise. Start with battery operation of the rig, and no antennas connected, and add components one at a time to determine which is the source of noise. The author lives in a rural area with very low environmental noise levels, and can operate all HF bands with three computers running, and virtually no sign of interference anywhere.

A useful tip concerning computer-monitor QRM – if you change the monitor resolution (easy to do these days), you may find the QRM goes away, or at least moves somewhere else.

Operator Comfort
Locate the computer keyboard and screen in comfortable operating positions; remember, you may be in front of the computer screen for many hours. Place the keyboard away from the edge of the desk, so the arms can rest on the desk. Place the monitor at a distance and height where it is easy to see without getting eye strain, or back or neck pain. For most operators this will be on a shelf about 200mm above the desk, perhaps above some accessory operating equipment (not the transceiver). Those with bifocal glasses may prefer to position the monitor lower.

Remember to leave space for the mouse, which should be placed so that your whole arm can rest on the desk. An 'L' shaped desk is great, because it offers room to swivel around to fill out the log on a clear piece of desk top. Light the desk so the keyboard and log are lit, but the light does not fall on the monitor. A low-power fluorescent desk lamp below eye level works well.

Place the transceiver where you can readily tune it while watching the computer screen, since this is where you do the tuning. It might require some compromise so that you don't end up leaning on the keyboard as you tune the transceiver. Thus placing the transceiver directly beneath the monitor and behind the keyboard is not a good idea.

Operating Hints
Perhaps the most fundamental requirement of digital-mode operating is being able to type! Younger operators will find this natural, but older Amateurs brought up BC (before computers were widely used) may have to learn from scratch. It is a good idea to use a typing-tutor program, which not only gives typing practice, but assesses and grades your performance, so you can see how you are improving. Many Amateurs, like the author, are hunt-and-peck typists, and will likely tell you they wish they had learned to touch type correctly.

Some Morse tutor software will also be helpful – sending Morse for you to type into the computer. With this type of software you can brush up on your Morse and your typing at the same time!

Typing Tricks
For those who are slow typists for one reason or another, or are just learning to type, there are digital modes that can be used where your slow typing won't show you up. SSTV is obvious, but some of the slower DX modes like Hellschreiber or experimental modes like Throb also have minimal typing

requirements. Hell operates at 20 WPM, or 10 WPM in DX mode, which is easy enough. These modes are generally operated without punctuation and in upper case only, so are much easier to type.

The best trick for slow typists is to make use of the type-ahead buffer. Most programs will let you type while you are receiving, and so long as you keep an eye on what the other station is saying, you can just about type slowly and continuously through his over as well as your own. When you start transmitting, just keep typing, and eventually the transmitter will catch up with you – hopefully close to the end of the over. The use of the buffer just about doubles your effective speed!

In many programs you can set up 'canned' messages, which also reduce the typing requirements. These are a good labour-saving idea, especially for procedural text like 'G0PFG de LA3KE pse K', and for standard information such as name and location. Other canned messages, such as a list of equipment, should be used very sparingly. Unfortunately it is becoming increasingly common for QSOs to consist of little more than an exchange of canned messages, which discerning operators should deplore. Mind you, many DX SSB contacts are little better!

Managing a QSO

This section is about managing 'chat' type QSOs [5]. Most digital-mode QSOs use a procedure and language little different from a Morse contact, except perhaps a little less terse. When you are getting started, it is a good idea to watch other QSOs for a while, to get some idea of the procedure. You'll also quickly recognise different styles, some you'll like and some you might not. You'll also recognise that the procedure is usually much the same for all the chat modes.

A CQ call consists of two or three lines of text, maybe with an invitation line added:

```
CQ CQ de G0PFG G0PFG G0PFG
CQ CQ de G0PFG G0PFG G0PFG
CQ CQ de G0PFG Cambs UK pse K
```

'K' is used to mean 'over', 'SK' to mean 'signing clear', while 'GA' (go ahead), 'R' (all understood), 'BK' (break) and other common Morse abbreviations are also widely used. KN means 'over, but no stations may break-in', which is not very friendly, and to be discouraged. Q codes such as QTH, QRL, QRM, QRN, QSO, QRX, QRZ and QSY can be used.

Many of the modern programs allow entry of callsigns for the stations that you call, for use with standard messages (often called 'macros'), which include the callsigns automatically. For example, in Nino IZ8BLY's excellent software, setting up a macro such as '$OTHER de $QRZ name is $NAME K' will make replying to another station's call very easy. Let's say you hear GM4XYZ calling CQ. You enter his callsign into the software, and when he stops transmitting, push just the one button for the necessary macro and you will automatically transmit:

```
GM4XYZ de G0PFG name is Fred K
```

The QSO will begin! It is usual to use a callsign exchange only at the end of each over, and if you set the callsign in the software as described, another macro '$OTHER de $QRZ K' will do the trick. It is usual for the first over or two between strangers to be fairly standard and stylised, with exchanges of name, location and signal reports.

After the first two exchanges, the QSO becomes less stylised, and, if the QSO lasts that long, more personal information is exchanged and questions asked and answered.

Depending on the mode, it may not be necessary to repeat text in order to make your meaning clear. (This depends on conditions and whether an error-correcting mode is in use). For example, under good conditions with a reliable mode such as Hell or MFSK16, just send your name and signal report once, and twice at the very most when conditions are poor. It is a good idea to give or repeat the name in capitals to make sure the other station sees it clearly:

```
OK John. Good sig hr, RST 549. Name is
Fred FRED and QTH is St Ives ST IVES.
```

Modes with no error correction (RTTY, PSK31) are an exception, and it is wise to send the important details several times if conditions are poor. Base your estimate of conditions at the other station on your own receiving conditions, then modify it based on what the operator reports.

Some modes (eg. RTTY, Throb and Hellschreiber using the traditional font) have a limited character set and no lower case capability, so it is accepted that text will be in capitals. As with packet radio and Internet e-mail however, if the mode has lower case, you should use it for preference, for two reasons:

♦ Upper case can be considered to be SHOUTING!
♦ The mode may transmit more efficiently and quickly in lower case.

This latter point is true of the varicode modes, PSK31 and MFSK16.

Accurate Tuning

Some digital modes, for example RTTY, Feld-Hell, SSTV or MT63, are not too fussy about tuning accuracy (well, no more so than SSB or Morse), but in general the digital modes need to be tuned very accurately, and some modes require extremely careful tuning and accurate netting. In fact, if you are not correctly netted to a station before replying to a call, the station may not come back to you, or may send 'QRZ?'. It may then take several tries to start a QSO.

The modes requiring the most accurate tuning invariably provide good tuning displays, automatic frequency control (AFC), netting, and precise tuning actually in the software. It is important to use this tuning facility, which can usually step 1Hz at a time, rather than the tuning on the transceiver, which can usually only tune 10Hz or 100Hz at a time. Using RIT to fine tune the signal is definitely not the thing to do.

For the Real Beginner

Before your first-ever digital QSO, it's a good idea to visit another operator's shack, to watch what happens and ask questions. Next, become familiar with reception, and then test the setup and software by transmitting into a dummy load until you are familiar with operating the equipment. Then arrange a contact with some local who has experience with digital modes, can send a good solid signal, and is able to critically assess your signal. Use a band that gives fairly good reception, and use a band segment that allows for initial SSB contact (and allows you to return to SSB if things go wrong). The 80m-band SSB segment during the day is a good place to start. Make sure the frequency you select permits SSB and digital-mode operation.

To make things easy, make contact in SSB first. This will make tuning a little easier. Make sure the RIT is off. Check that your digital-mode connections are made and ready to go. When you have made contact, and wish to switch to a digital mode, remember that the other operator may require some time to change over – most operators will need to unplug or disable the microphone, and connect up the computer. Connect up your own cables and give it a try.

Some modes are easier to get started with than others. Hellschreiber (Feld-Hell) is especially suitable, while RTTY is also reasonably easy, although be aware that you need to have the correct sideband selected or you will copy nothing. SSTV is also fairly easy if you can manage the software. After you've mastered these, tackle the more challenging modes, perhaps trying PSK31 next. You might wish to start with just one 'multimode' program so you have less software to learn.

Choosing Software

Selecting software is a bit like choosing a new hi-fi or a bottle of wine, as there are many choices and everyone has different tastes. However, broadly speaking, it is a good idea to use as few packages as possible to achieve all the modes you wish to run. There are numerous advantages to multi-mode software – convenience of course, the easy familiarity that comes with needing to learn only one set of commands, and speed of mode change.

Some of the newer multi-mode software, such as MIXW 2 by Nick UT2UZ and Denis UU9JDR (**Fig 6.11**), offers almost every commonly-used mode. However, the discerning specialist may prefer to use software that concentrates on just one mode. For example, IZ8BLY Hellschreiber is the undisputed software of choice for Hell, while MMTTY by Mako JE3HHT is widely respected as the RTTY specialist's program. There are dozens of software choices for PSK-31.

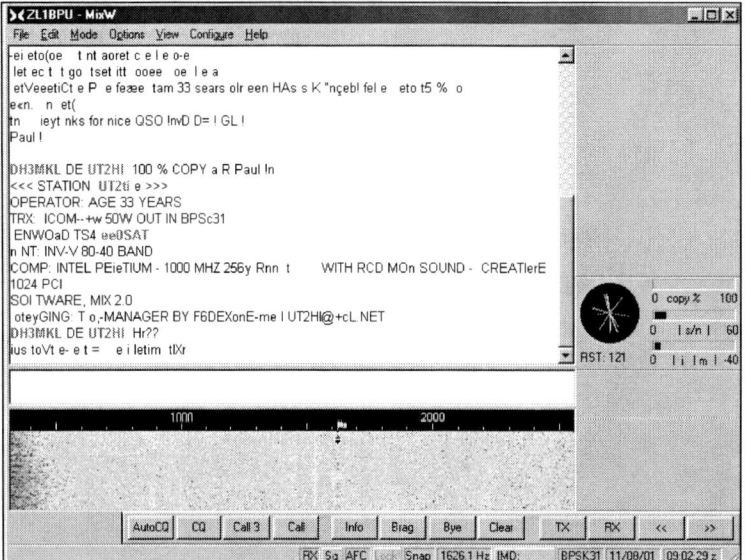

Fig 6.11: MIXW 2 by Nick UT2UZ and Denis UU9JDR

Amateurs who prefer logging on the computer will enjoy using the many programs that now offer automatic logging. Unfortunately there are many different formats, which makes keeping a single multi-mode log difficult unless a single program is used for all modes. Some digital-mode programs (for example MIXW) are able to log contacts for modes that the software does not operate – such as SSB or FM.

It is a good idea to check first that the software you are interested in will run on your computer! The documentation will normally indicate what processor speed should be used, what type of sound card, and how much memory is required. As DSP techniques improve, more modes are added, and more processing is added, and the minimum system requirements continue to increase.

It is possible to run the simpler modes on low-specification computers. For example, the original PSK31 software by Peter G3PLX operates quite satisfactorily on a 486 computer at 33 MHz. There are also versions of RTTY for DOS PC by Rob ZL2AKM, and Hellschreiber by Lionel G3PPT and Sigfus LA0BX that operate on a 486 computer.

Almost all the sound-card based software used for digital modes is free, and widely available on the Internet. The programs with highest performance (and most available modes) tend to be commercial software, although frequently they are fully operational for a limited period without registration.

Without doubt, the best place to look for software is the Internet. Appendix B lists the best sites from which to obtain both free and commercial software.

Special Operating Considerations

Most of this manual is about conventional digital modes operated from conventional suburban home amateur stations. There are, however, other special circumstances that digital-mode operators find themselves in, which need to be considered.

Operating Low Power

QRP or low-power operating is one of the most challenging and interesting aspects of digital modes. Some of the modes are especially suited to QRP.

Unfortunately many traditional QRP transmitters are CW only, so are not suited to sound-card digital-mode operation. As a result, most QRP digital-mode operation involves using a 100W SSB rig with the power wound down, or one of the lower powered SSB transceivers. Fortunately power control using a sound card is very easy.

Special digital-mode QRP transceivers are now available, both built up and as kitsets. One of the best known is the 'Warbler' family by Dave Benson NN1G (**Fig 6.12**) [6]. The 80m Warbler designed by Dave and sold as a kit by the New Jersey QRP Club is inexpensive and small enough to fit in your pocket.

Morse is obviously the most popular QRP mode, and PSK31 is

Fig 6.12: The NJQRP 80m Warbler (Photo: NJQRP)

deservedly popular as well, as it offers very high sensitivity. Some of the other modes available are also excellent QRP modes. MFSK16 is especially good for 80m QRP, offering ranges of 2000 km and more with 5 Watts, and several hundred km on 100mW or less. PSK-Hell is another mode with good performance on noisy bands. Feld-Hell is a good choice for CW-only transmitters.

On the higher bands, PSK31, MFSK16 and PSK-Hell are all good choices, with MFSK16 being the best for long path. The author has logged many EU-ZL contacts on 20m and 17m using 5W PSK-Hell or MFSK16.

When portable or field-station operation is being considered, Hellschreiber is a very good choice, as the duty cycle of the Feld-Hell transmitter is very low (about 20%), resulting is a signal with very good 'punch' at low-average power-supply current. Morse is also of course a good choice. Most of the other digital modes operate the transmitter at near full power, and obviously consume more of the precious battery reserves.

Stealth Operating

Apartment and flat dwellers are increasingly being restricted by high interference levels and the impact of their operating on the neighbours' TV, phone and stereo equipment. Newer housing developments frequently have restrictions placed on the owners, preventing erection of visible antennas.

The usual approach to stealth operation is to use hidden antennas – antennas in attics, along fences or otherwise disguised. Such antennas tend to be noisy and inefficient, so stealth stations would do well to embrace digital-mode operation to achieve a better level of performance.

The interference factor to stealth operation – ensuring that the neighbours never know that a transmitting station is next door because they observe no interference – is best overcome by following two approaches. First, operate low power, to minimise the risk of interference through overloading. Second, choose a mode with reasonably constant transmitter power. Morse, SSB and Hell are not recommended! The best modes are RTTY and MFSK16, which are true constant amplitude modes. FM and SSTV are also constant amplitude modes. PSK31 and PSK-Hell may also be suitable as the transmitter power is reasonably constant.

Moon Bounce

Not something that every amateur enjoys, EME operation is certainly for the expert enthusiast, as the equipment is expensive and very specialised. Most operation takes place on 70 cm and 23 cm bands, using slow Morse, often computer transmitted and hand received, enhanced by special digital filters. Special shortened signal-reporting techniques are used. There is now excellent FFT 'waterfall' software designed for recovering weak carriers 20dB or more below the noise level. A good example is Spectran by Alberto I2PHD, which has an FFT display, filters and a de-noiser.

Digital-mode EME operation is now being considered. Most existing digital modes are not suitable due to stability difficulties on UHF, combined with the unique fading characteristics of moon-bounce signals. Some of the better contenders are slow Hellschreiber (eg 1/8 speed), and VSKCW (dots and dashes the same length, but on slightly different frequencies). There is also interest in fast specialised MFSK modes such as FSK441.

Satellite Operation

Digital operation through satellites (apart from those that operate packet modes) is quite difficult compared with normal terrestrial operating because of strong Doppler shift and long delays. PSK31 and other really narrow modes are especially difficult to use. A number of operators report good results with Feld-Hell, which is probably the least fussy of the more widely available modes for tuning accuracy.

Fig 6.13: Aurora DX on 2m (Picture: Rex Moncur)

Aurora and Meteor Scatter

Aurora operation is a little like bouncing a signal off a fuzzy mirror. The signals returned are very noisy (strong amplitude variations), with highly distorted phase (**Fig 6.13**). Most digital modes are not suitable. Amplitude-keyed modes like Morse and Feld-Hell are effective. Rex VK7MO reports QSOs on 6m and 2m over 2000km paths on numerous occasions, using slow Hellschreiber (1/8 speed).

Meteor scatter is popular with VHF operators, but requires quite different operating techniques – in fact virtually opposite to Aurora. The 'pings' (very short reflections from tiny meteorites) and 'burns' (longer reflections from ionised particles left in the upper atmosphere) cause strong signals of very transient duration, in a highly unpredictable manner. Six metres is the most effective band, but meteorite pings can be observed readily from about 10MHz to 500MHz and above.

During the more active meteorite events, which occur on several occasions annually as the earth passes through known areas of dust such as the tail of a comet, meteorite events are so numerous that even SSB communication is possible, but at other times, communication depends on passing as much information as possible during the random and very transient reflection events. Thus a very high-speed mode is called for.

Another problem facing MS operators is that until an event occurs there is no signal to tune to, so there is the need to know the other station's transmitted frequency with considerable precision, not readily achieved at VHF.

In the past, high-speed Morse was popular for MS, using tape-recorded transmissions and some means of slowing the recorded reception down for reception. This technique is slow and cumbersome. More recently digital recording and digital processing have allowed the recordings to be searched for pings and played back or displayed for manual reception.

New techniques for automatic reception of MS signals are now evolving. One of the problems with meteorite pings is very strong Doppler shift, since the particles burn up in the atmosphere at very high speed. This makes normal narrowband techniques impossible to use.

Experimenters report good success with high-speed Hellschreiber (available in the IZ8BLY software) at five or nine times normal speed, up to 180 WPM. This mode has the merit of requiring no synchronism, easy interpretation of partially received characters, and good tolerance of Doppler and frequency errors.

FSK441 is an interesting high-speed MFSK mode developed especially for MS by Joe Taylor K1JT. This mode uses four tones spaced 441 Hz apart, and

Fig 6.14: The WSJT meteor scatter software in action

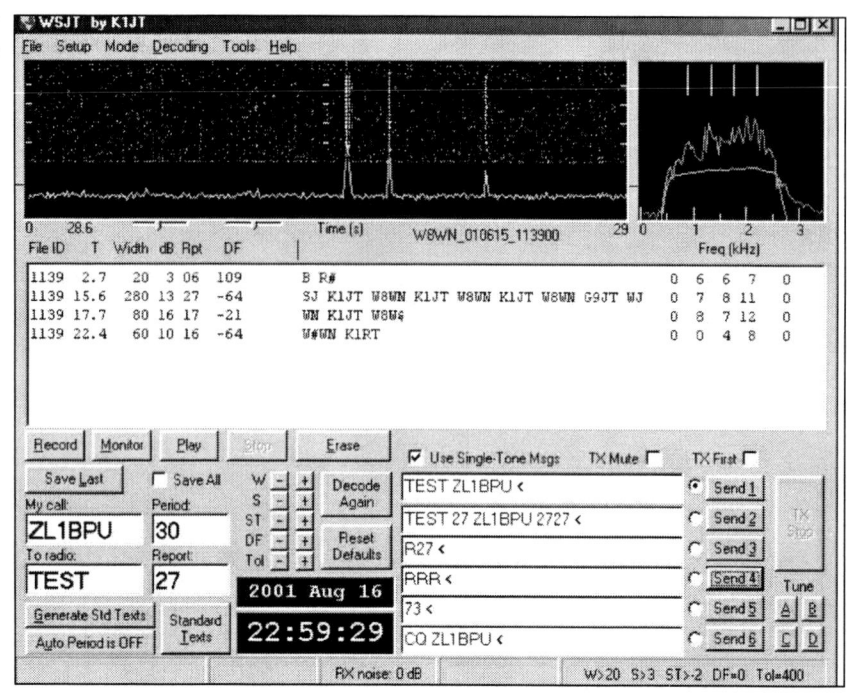

operates at 441 baud (about 880 WPM). The WSJT software (**Fig 6.14**) transmits short phrases over and over, then switches to receive and records for a fixed period, then analyses the recording and automatically displays the text received.

References

[1] The schematic of a typical interface of this type can be found at http://www.qsl.net/zl1bpu/FUZZY/media/schem.gif, and the parts list at http://www.qsl.net/zl1bpu/FUZZY/media/partlist.gif.
[2] See later under 'Setting Up'.
[3] See Appendix B for suggestions.
[4] An open-frame or unshielded antenna tuner is often overlooked as a potential point of noise entry.
[5] The procedure for ARQ type connected modes is sometimes quite different, and where necessary is treated in the chapter related to the particular mode. Better still, read the MCU Operator's Manual.
[6] See Appendix C for web site references.

7

AmTOR

In this chapter
- AmTOR Summary
- Error correction
- AmTOR Mode A
- AmTOR Mode B
- Operating AmTOR
- Performance

AMTOR IS BASED on a commercial system developed in the late 1950s for long distance fixed and marine services. The system was called TOR (Teletype Over Radio), and used digital techniques to control and correct errors for conventional teleprinters. The marine version, SiTOR (Sea TOR or Simplex TOR) was widely used, and is still in use today for ship-to-shore services.

SiTOR was designed to be an exact match in speed with 50 baud teleprinters. The earliest equipment was constructed using mostly digital logic, and was used to translate to and from ITA2 for direct use with teleprinters. There is a 1:1 correspondence between the SiTOR characters and those of the teleprinter, so conversion was relatively straightforward. Operation was therefore also straightforward,

AmTOR Summary

Symbol rate	100 baud
Typing speed	6.6 CPS (66 WPM)
ITU-R description	400HF1B
Bandwidth	400Hz
Modulation	2-FSK
Average power	50% (ARQ), 100% (FEC)
Protocol	Synchronous connected ARQ CCIR 476-4 (Mode A)
	Synchronous unconnected FEC CCIR 476-4 (Mode B)
Character set	Moore 4:3

since the operators were familiar with the same teleprinters and terminal equipment in RTTY service.

Peter Martinez G3PLX first evaluated the system for amateur applications, and published a series of articles [1] including a description of suitable equipment for this mode in 1979, at a time when microprocessors were first becoming popular for home projects. The following year Peter offered a microprocessor construction project [2] which became the forerunner of many AmTOR equipped MCUs. AmTOR (Amateur TOR) was the first Amateur mode to use error correction, and was quite widely used in the 1980s and early 1990s.

The error correction in AmTOR is modest by modern standards, but AmTOR revolutionised RTTY links and especially automated systems, making bulletin board and message-forwarding systems practical for the first time. The parity detection system (see below) is the weak link in the system. Errors that occur when two bits in a character are damaged by noise will not be detected.

AmTOR is in fact two independent but related modes. One is an ARQ mode (AmTOR A) intended for connected operation between two stations. The other, AmTOR B, is an FEC broadcast or net mode, designed so that one station can transmit to many recipients. These two modes correspond exactly to SiTOR A and SiTOR B, as described in CCIR 476-1 to -4 and CCIR 625 [3]. A third mode often provided by Amateur equipment, but not envisaged for commercial service, allows casual (but unconnected and uncorrected) monitoring of the sending station in an AmTOR A link. This listen-only operation is called AmTOR mode L. Performance in this mode is modest, all the ARQ repeats will be printed, none of the errors are corrected, and if the sending station is much weaker than the receiving station the reception may be patchy due to receiver desensing. However, Mode L provides the only way to monitor or identify ARQ stations in QSO.

AmTOR modes A and B share the same baud rate (100 baud), the same character set, the Moore code, and therefore the same parity-checking technique. The modes differ in how the errors are handled. In both cases, the signal is transmitted as 170Hz shift FSK, and the signal bandwidth is about 400Hz.

Fig 7.1 shows the audio spectrum of a real AmTOR transmission. There are many similar images in following chapters, and are all made with the same equipment and the same settings [4].

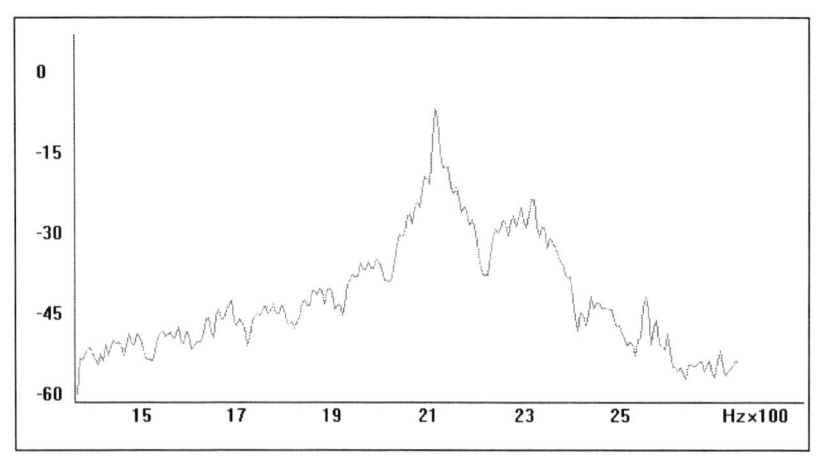

Fig 7.1: Spectrum of AmTOR

Error Correction

The error-correction technique used by AmTOR is, by modern standards, relatively weak, but it is simple to use, reasonably fast, and at the time of its development offered a level of performance well in advance of the uncorrected alternatives. Of course the system was devised before the days of microprocessors, so the error-correction algorithm needed to be simple enough to implement in hardware.

The Moore code is a systematic parity system, meaning that the transmitter generates, and the receiver checks for, a predetermined number of zero and one data bits per character. Each Moore code character has been expanded from the ITA2 alphabet into a seven-bit character code, where of the potential 128 possible combinations, only those with four '1' bits and three '0' bits are used [5]. Thus at the receiver, if the character does not have the correct number of the requisite bits, there must be an error.

The simplest-possible parity system, where the number of ones or zeros is kept constant by the addition of a single parity bit to each seven-bit ASCII character, is widely used on direct-wire-connected computer equipment, and is often sufficient in that environment. However, where noise is present, even the Moore code, with effectively two bits of added parity, is barely adequate. The problem arises when two bits in a Moore character are corrupted, and yet the parity conditions are still met. It is then not possible to tell if the received character with correct parity is indeed what was sent.

Thus the ARQ mode will still print some errors, since there are some it cannot detect. The FEC mode has an extra opportunity to detect errors, since each character is sent twice, and the two instances can be compared. Mode B has no opportunity to correct errors at all, if the two versions of a character are different and yet the parity is in both cases correct. In this case of uncertainty (as also when both versions of the character fail the parity check), a space or underscore '_' is printed instead. Very often printing an obvious space or underscore to indicate the presence of an unknown letter will give better readability than that which includes printing of unrecognised errors.

AmTOR Mode A

In mode A, the ARQ mode, the master (information transmitting) station transmits for 210ms, and then listens for 240ms, and then transmits again. The mode has a highly distinctive chirp-chirp-chirp sound at this fast rate. The slave station sends a much shorter 70ms chip-chip-chip signal at the same rate. When the stations are connected, and both within range, a distinctive chirp-chip-chirp-chip sound can be observed.

The baud rate used is 100 baud, one bit per symbol, and so each character (which contains seven Moore code bits) takes 70ms to transmit. Each 'chirp' thus transmits three characters. The slave (information receiving) station checks that the received data are correct, and in the remaining 240ms window transmits back just one character, indicating whether the entire chirp was received correctly or not. The time at which this character is received varies considerably, depending on the propagation path. The delay of this response includes the time to transmit and receive the message, check the parity, calculate the response, and the time to transmit and receive the response. Imagine the range between two stations is 20,000km (60ms); the total response delay resulting at the master station will easily exceed 190ms (two path delays plus the time for the response character). For this reason AmTOR mode A can be difficult to operate

DIGITAL MODES FOR ALL OCCASIONS

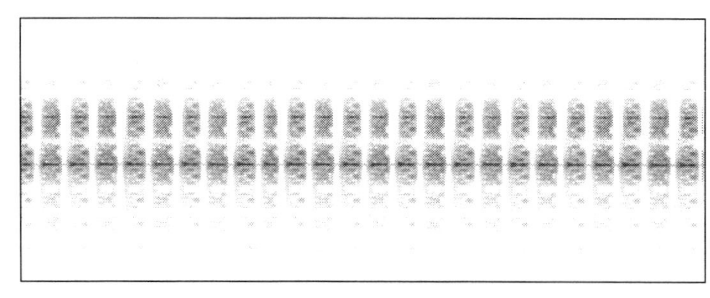

Fig 7.2: Spectrogram of AmTOR mode A

at very long range, especially on long path. To achieve reliable operation over long distances the timing of the response needs to be adjusted earlier, and the setting used will probably be too early for short-distance contacts.

The ARQ response to each three-letter packet alternates between two of the special control characters, known as Idle Signal a and Idle Signal b. When the data received check out as correct, the control-character returned changes, but is not changed from the previous response if the data contained an error (ie the parity of one or more characters failed). Thus the master continues with the next block of three characters only while the response characters alternate.

In the spectrogram (**Fig 7.2**) (of one station calling another) the gaps between the transmissions can be clearly seen. The calling process uses a sequence of letters (called the 'selcall', or selective call) to attract the attention of the other station. The usual convention is for Amateur selcalls to consist of the first callsign letter and the last three letters, although any four letters will in fact suffice. A CQ call can be made using CQCQ as the selcall, but it was general practice to use mode B for CQ calls, as the transmitting station is not identified during the mode A CQ call.

A special sequence of messages is used to establish connection, to change stations from master to slave (like changing from transmit to receive), and to disconnect the link.

AmTOR Mode B

This is the FEC mode, generally used for sending bulletins and calling CQ. The transmitting station sends each character twice, five characters apart. This delay provides a simple interleaving capability that helps counter burst noise (eg lightning). The receiving software will print each character as the second version arrives, provided at least one version is correct. The software will generally print a blank or underscore '_' indicating an error, if the two versions of the character do not agree, or both fail the 4:3 parity check.

Each character is printed after a five-character delay. As in mode A, each character consists of seven bits, each 10ms long, so the delay between the first and second transmission of each character is 350ms. Thus the receiver prints at least this much later than the transmitted signal is received. This is called the 'latency', and is solely due to the interleaver. The AmTOR mode B spectrogram is shown in **Fig 7.3**.

Fig7.3: The AmTOR mode B spectrogram

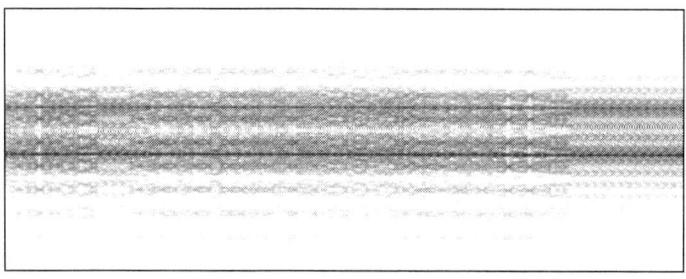

There are commercial selective-calling versions of mode B, known as SelFEC and NavTEX. NavTEX marine broadcasts on MF are readily copied with

Amateur AmTOR equipment. AmTOR mode B sounds very similar to RTTY, but has a faster, more steady, bdrr-bdrr-bdrr sound.

Operating AmTOR

Most HF MCUs offer AmTOR, although it is now little used, having been widely superseded by PacTOR. Software exists for DOS PC and standard RTTY TUs [6], for DOS PC and 'Hamcomm' interfaces, and for a few specialised DSP units. There is no known mode A software for Windows sound card, probably due to timing difficulties and lack of demand, but a very effective implementation of mode B is offered in the MIXW 2.0 multi-mode program by Nick Fedoseev UT2UZ, and Denis Nechitailov UU9JDR [7].

The commands and procedure used for AmTOR A varies considerably with the equipment used, although the actual protocol on air is the same. A request to change direction is made by the master station by sending the three-character group 'FIGS +?', and the actual changeover is then initiated at the former Slave station via special control characters. The slave can also initiate change over directly. In bulletin-board service, the host server normally controls all the changeovers for you. Because equipment differs widely, it is best to consult the documentation supplied with the equipment used for operating advice [8].

AmTOR B operation is very simple, being no different in practice from RTTY or Hellschreiber, and very similar to Morse.

Performance

By modern standards, AmTOR is quite slow. In ARQ mode, the potential 6.6 CPS (66 WPM) is rarely reached, and when conditions are poor the link is continually lost and re-established, and little data are transferred. Because AmTOR uses 100 baud FSK, it suffers severely from multi-path propagation effects. The error correction by modern standards is simple, but not very robust. AmTOR ARQ operation has been completely replaced by improved modes such as PacTOR. The FEC mode B alternative is also slow, but is still very effective for news bulletins where modest FEC is sufficient.

References

[1] For example, 'AmTOR, an mproved radioteletype system, using a microprocessor', J.P. Martinez, *Radio Communication*, August 1979.
[2] 'AmTOR, the easy way', J.P. Martinez, *Radio Communication*, June/July 1980.
[3] CCIR 476-1 'Direct Printing Telegraph Equipment in the Maritime Mobile Service'. Annex I describes all the characteristics of ARQ, FEC, the Moore code and the protocol.
[4] The settings used throughout the book are 5.5kHz sample rate, FFT size 1024, Scope mode, dwell 20 ms, spectrum average = 128. The software is Spectrogram V 4.2.6.4.
[5] The Moore code is listed in Appendix B.
[6] For example 'BMK-Multy' by Mike Kerry G4BMK.
[7] http://tav.kiev.ua/~nick/mixw/mixw.htm (latest release).
[8] A simple but detailed explanation can be found in 'Introduction to and the Operation of AmTOR', Phil Anderson W0XI, Kantronics Inc, 1983.

8

Clover

In this chapter
- Clover II modulation
- Error correction
- Performance
- Clover 2000
- Operating clover

CLOVER IS AN advanced communications mode, or rather, a family of modes, based on a special waveform designed by the late Ray Petit W7GHM in the early 1990s. The most widely-used version is Clover II. All variants are commercial systems, and are only available in equipment made by or under licence to Hal Communications [1].

Clover uses a four-carrier orthogonal frequency division multiplex (OFDM) technique [2], where different pulses are sent on each of the four frequencies in turn. The tones are 125Hz apart. There is a pulse on each frequency in every symbol period, thus it is possible to send many bits of data per symbol, as the phase and amplitude of each of these pulses can be varied to send the data. The symbol period is 32ms, so each pulse is centred 8ms after the one before. There is some overlap between pulses, so that while the transmitter power is relatively constant, the transmitter must operate linearly. The pulses are carefully shaped to minimise

Clover Summary

Symbol rate	31.25 baud
Typing speed	Approx 3–50 CPS (30–500 WPM)
ITU-R description	500HJ2DEN or 500HJ2BEN
Bandwidth	500Hz
Modulation	8, 4, 2-PSK, 4, 2-ASK
Average power	50%
Protocol	Synchronous connected ARQ with Reed-Solomon FEC coding
	Synchronous unconnected FEC with Reed-Solomon FEC coding
Character set	ITA-5 ASCII (any 8-bit code)

DIGITAL MODES FOR ALL OCCASIONS

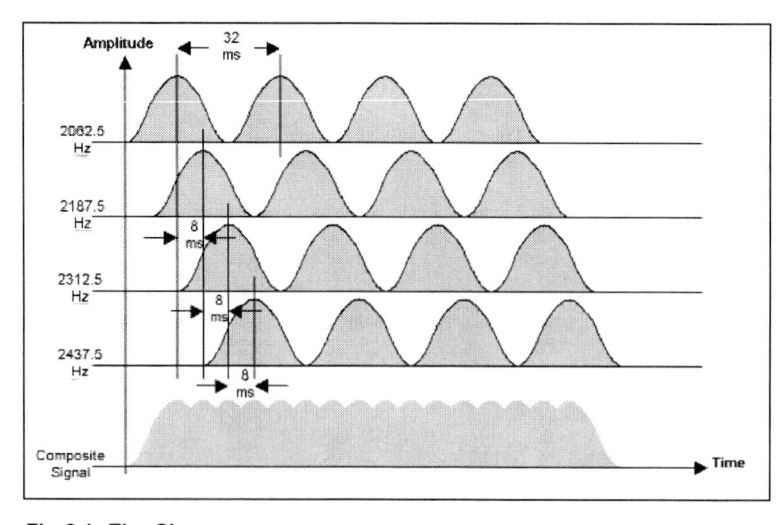

Fig 8.1: The Clover pulse sequence

Fig 8.2: The Clover spectrum

bandwidth. The Clover II transmission is exactly 500Hz wide, and has a prrrrrrt-prrrrrrt-prrrrrrt sound, rather like a cricket chirping (**Fig 8.1**) [3].

The spectrum is always the same, no matter what modulation scheme is used. Looking at **Fig 8.2**, the signal is very clean with four obvious peaks and steep sides. Clover can be used for bulletin-board systems, but most operation at present seems to be for regularly-scheduled DX rag-chew contacts. Clover is not as widely used now as during the mid 1990s.

There are numerous modulation techniques used, and the protocol allows these to be changed automatically as conditions dictate. Although the modulation can change, and with it the data rate, the signal symbol rate (31.25 baud) and the signal bandwidth, are always the same. Clover was designed to be tolerant of multi-path reception, as is evident from the low symbol rate and, by offering a number of different modulations and protocols, can adapt to a wide range of conditions. Whether the software used is able to choose the optimum protocol for the conditions is somewhat a matter of conjecture! The mode also includes error correction, and is also very tolerant of burst noise.

Clover II Modulation

The five Clover II modulation modes are summarized opposite. The faster modes require very stable conditions, and all modes require high transmitter and receiver stability. The performance of even the BPSM mode is good because each of the four carriers is transmitting one bit per symbol. The 2DPSM mode uses two tones for each data bit. The wide range of modes available allow Clover to operate to advantage over a wide range of conditions.

The ARQ protocol used by Clover is quite unusual and allows what amounts to, or at least simulates, a full-duplex conversation, provided that the typing rate is modest.

The ARQ connected mode is controlled by the exchange of a short Clover control block (CCB), which transmits such information as callsigns, mode-control data and signal statistics, but can also carry a small amount of keyboard

Mode	Description	Data rate (bps)	Protocol
16P4A	16-PSK 4-ASK	750	ARQ, FEC
8P2A	8-PSK, 2-ASK	500	ARQ, FEC
8PSM	8-PSK	375	ARQ, FEC
QPSM	4-PSK	250	ARQ, FEC
BPSM	2-PSK	125	ARQ, FEC
2DPSM	2-PSK diversity	62.5	FEC

text. When the link is established and the typing rate is low, only CCB blocks are exchanged.

As data waiting to be transmitted increase, larger blocks called 'data blocks' are added, so that when one station is typing fast or sending a file, the other station can still send slowly, apparently at the same time. When data are waiting to be sent at both ends, both stations send data blocks as well as the short CCBs, and the channel is used at about the same rate in each direction. See Fig 8.3 for the Clover ARQ spectrum.

Error Correction

Clover uses a Reed-Solomon block FEC system in all modes. Errors are detected on 8-bit groups of data (bytes). Using the ARQ protocol, this allows most errors to be corrected without the need to request a repeat, which improves the operating efficiency.

The coding is not performed in the modem, but is performed by the host application, so the strength can be adapted to suit the conditions. Block lengths of 17, 51, 85 and 255 characters are used, along with four different coding efficiencies.

The FEC mode also has the benefit of the Reed-Solomon ECC, and while there is no ARQ correction, and no two-way exchange of control block path information, the transmitting station can at least change the FEC strength, and the receiver will automatically detect which ECC mode is in use. FEC mode is of course useful for CQ calls and net operation. See **Fig 8.4** for the Clover FEC spectrogram.

(Top) Fig 8.3: The Clover ARQ specrogram

(Bottom) Fig 8.4: The Clover FEC spectrogram

Performance

Due to the adaptive mode and error-correction techniques used, Clover II performs very well in most conditions, and is especially good where multipath is a problem; it is therefore frequently used by its fans for

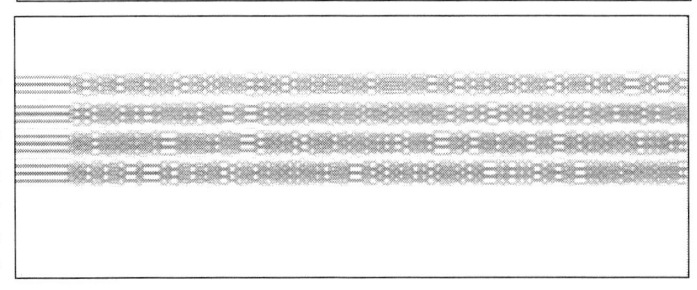

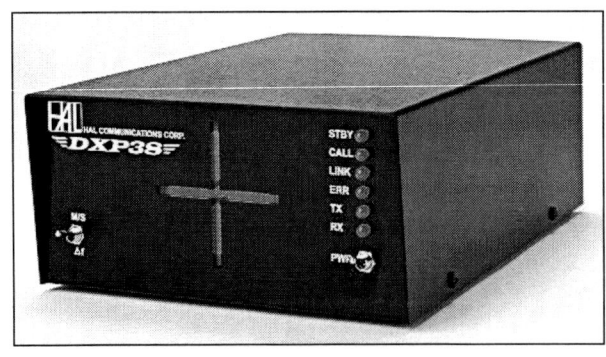

Fig 8.5: The HAL DSP38 DSP modem

long-distance DX, especially long path. When signals are very weak, the performance (quoted at about 20bps for 0dB S/N) is about twice that of AmTOR. This is only part of the story, because the transmitted signal skirts are better controlled, the protocol supports bi-directional ASCII text, the multi-path performance is far better, and the error correction is much more robust. In addition, as conditions improve, the speed increases dramatically with the adaptive technology, approaching 60bps. AmTOR of course just plugs along at the same slow speed. Another major advantage is the signal-link information contained in the CCB, allowing the user to monitor the modulation mode, path phase dispersion, signal-frequency error, the error-correction performance and signal-to-noise ratio of the signals at both ends of the link.

Comparison of Clover with other modes is the subject of much discussion, and although such comparisons can only ever be subjective it is generally held that it out-performs PacTOR but not PacTOR II.

Clover 2000

Intended as a much higher-performance commercial alternative to Clover II, and optimised for file transfer, Clover 2000 occupies 2000Hz bandwidth, but otherwise is similar in concept, using eight tones rather than four [4]. It has excellent multi-path and selective fading performance. Clover 2000 is available in only a very few modems intended for commercial use, and has not been widely used in Amateur circles. Signals of this bandwidth are not popular on the ham bands!

Operating Clover

Clover II is very complex to generate and to receive, so that only commercial DSP-based equipment can operate this mode, for example the HAL P-38 and DSP-4100 (**Fig 8.5**). However, the software provided with the equipment is reasonably straightforward to use, and with practice, operation is no more difficult than other modes. The documentation recommends that you tune in the signal slowly and accurately, using the tuning display on the computer screen, then leave the tuning control alone, and preferably lock the dial.

Probably the least-expensive way to get started with Clover is to locate a second-hand P38 unit. This is a full-length PC plug-in card, and will operate in even the most modest DOS PC with a 286 or better processor. It also generally works quite nicely in a Windows 95 or 98 'DOS box'.

References

[1] http://www.halcomm.com/.
[2] 'Orthogonal' means that the four carriers can carry completely independent data without interaction.
[3] See Clover II Waveform & Protocol, Hal Communications Corp E2006, 1997.
[4] http://www.halcomm.com/clover2000.htm.

9

Hellschreiber modes

In this chapter
- History of Hellschreiber
- How Hell works
- Feld-Hell
- Multi-tone Hell
- PSK Hell
- Other Hell modes

THE HISTORY OF Hellschreiber is long and illustrious, but surprisingly little known and sparsely documented [1]. Developed by Dr Rudolf Hell in 1927, the Hellschreiber ('bright writer') was devised as a simple means of distributing text from central press offices to newspapers. At a time when teleprinters were complex, expensive, and not particularly reliable, Hell conceived a simple, quiet machine with very few moving parts, and a system incorporating numerous technical advances that made it considerably ahead of its time.

Fig 9.1: The Siemens A2, 1944 (Photo: Dick Rollema)

The machines were soon used widely on land lines for press use, and for the same purpose via radio transmissions from about 1930. Military versions emerged around 1937, and were used by German troops during the Spanish Civil War. A landmark technical paper 'Die Entwicklung des Hell-Schreibers' ('The development of Hell writers'), was published by Rudolf Hell in 1940 [2], the same year that the Siemens A2 Feldfernschreiber (**Fig 9.1**) [3] (used widely throughout World War II and later), was described in detail [4].

Reuters and other press agencies continued to use Hellschreiber for world-wide press transmission until the 1960s, when it was replaced by the Telex service. The press machines operated at 245 baud.

Fig 9.2: Sample from a 10,000km HF Hell QSO

Countries with ideographic languages, such as China and Korea, continued to use Hellschreiber variants until the mid 1990s, when they were finally replaced by FAX.

Amateur use of Hell started to grow in the 1970s, and reached a peak in 1999 with the introduction of high-performance DSP Windows software and more robust Hell-type modes. Much of the early enthusiasm can be attributed to the 'PA Hell Gang' formed in 1979, which still meets, and several articles in Amateur magazines [5]. Many original Hell and Siemens machines are still operational and used regularly, including the excellent example in Fig 9.1, owned by Dick, PA0SE.

Fig 9.2 shows a sample from a 10,000km HF Hell QSO.

The US Army RC-58B (**Fig 9.3**) system from the 1940s transmitted an FSK signal using optical scanning of hand-written text, using a marvellous hexagonal spinning prism, but the receiving system was virtually identical to the Hellschreiber. By virtue of the optical scanning, this equipment was able to retransmit received tapes. It was used in light armoured and communications support vehicles.

During World War Two, to overcome the shortage of press Hell receivers, the GPO manufactured their own copy of the Hell receiver. They were designed at Dollis Hill, and assembled by an adding-machine manufacturer! Similar machines were made in other countries, notably Japan, and there is a story that the late Stan Cook G5XB, a long-time Hell enthusiast, built a receiver from Meccano while working for the BBC during the war.

Fig 9.3: The Receiver-scanner unit of the US Army RC-58B (Picture: Marc Kulbacki)

In 1937 an alternative system, the seven-frequency system (**Fig 9.4**), was described, but although tested on a path from Algiers to Paris, it did not see commercial service [6]. Peter Martinez G3PLX developed a DSP-based system with a similar concept in 1997. Other systems with purely Amateur application have also been developed in the last few years, and will be described later.

Technically, the 1929 Hellschreiber was both enchantingly simple, and technically advanced. Signal bandwidth was managed by a clever technique that allowed 4ms image resolution while only requiring 8ms equivalent bandwidth; no synchronism was required, and the system was very noise immune. The system made the first known use of audio subcarriers for transmission, and the first use of electronic transmission and reception for digital modes. It

CHAPTER 9: HELLSCHREIBER MODES

used audio-tone autostart, was applicable to multiple character sets and ideographic languages, and was operated widely and successfully by radio some 12 years before RTTY. In later years there was also a page printer and even a machine for transmitting cheques between banks for signature verification.

At the time of writing, Rudolf Hell died recently at the age of 100 years. Apparently Dr Hell was quite interested in recent Amateur improvements to the technology he started so long ago.

Fig 9.4: An example of text from the 1937 French seven-frequency system

How Hell works

Hell is a fuzzy mode [7]. Thus there is no encoding of the data, and no timing or amplitude decisions made in the receiver. The older equipment was not fully compliant with all the fuzzy rules since it was not possible to print faint dots when the signal received was faint, as is now possible. However, the other rules were followed.

Transmitting Hell

Each character to be transmitted is scanned much like a miniature facsimile, usually with 14 dots per column and seven columns per character. In most cases the scanning was electromechanical rather than optical scanning of an image. By carefully controlling the definition of the font used, individual dots are never transmitted (the minimum number is two consecutive dots), so the bandwidth is much reduced. Each character typically has ten printable dots out of 14 per column, and five printable columns, the remaining four dots and two columns being white space. The characters are scanned vertically from the bottom left corner, then in columns left to right.

The original transmitters used a simple contact drum (one ring of contacts per character) or set of cams (one per character) to close contacts and operate an audio oscillator or CW transmitter. There were also machines which transmitted text translated from five-hole ITA2 punched paper tape. Newer systems use computers with digital memories and a choice of special fonts. These systems transmit carefully shaped minimum bandwidth dots (**Fig 9.5**) [8].

In Fig 9.5 the time scale is 5ms per division. Individual shaped dots are 8ms long; the space between the first two dots is 8ms. The space between the second dot and the following series of joined dots (like a dash) is 12ms, so although no dot is smaller than 8ms, the resolution is 4ms. The result is a signal with a bandwidth of about 350Hz.

Receiving Hell

The original receivers had a spinning helix (like a two-turn worm gear), rotating at column-scan speed. The helix had an inked roller running on it. The helix was positioned just above a slowly-moving paper strip. As an audio tone was detected by the receiver electronics, it caused a solenoid to pull in, moving a hammer bar attached to the solenoid armature to strike the paper from underneath, so that it touched the inked two-turn helix in two places. The motor operating the helix also slowly drew the paper tape along past the helix, using a slowly-

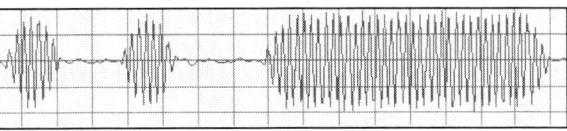

Fig 9.5: Oscilloscope sample of Hellschreiber dots

83

Fig 9.6: Text is easily readable with 0.2% speed error

rotating shaft and a pinch roller, similar to those in a tape recorder. The motor was speed governed and the speed was adjustable via the governor to ensure that text printed straight, or nearly so. In the wartime A2 machine speed control was also electronic!

The better modern receivers are able to provide very good speed accuracy, and simply print a dot when sufficient energy is received from the receiver, printing a grey-scale image where signal strength is represented as darkness of dots. In this way, the eye and brain are able to best interpret the received information. To replicate the two-turn worm so that the text prints twice, the receiver prints each received dot twice, one above the other, 14 dots apart, (see Figs 9.2 and 9.6) so that it does not matter what phase the receiving system has compared to the transmitter, there will always be one version of text readable, and in fact the text remains readable with a wide discrepancy between transmitter and receiver speeds (**Fig 9.6**).

In the modern DSP receivers the band-pass filters, detector, low-pass filter and printing are achieved by mathematical means. These systems also include automatic gain control (AGC) and software tuning of the receiver to the received tones.

Feld-Hell

The most widely used Hell mode today is still the mode used by the 1940 wartime machines. These transmit using on–off keying, at 17.5 columns per second, or 122.5 baud with seven dots per column. This results in a text speed of about 2.5 CPS (25 WPM). The name Feld-Hell comes from that of the original Siemens machine, the A2 'Feldschreiber' or field writer. Almost all users now operate Hell using sophisticated sound-card software, and contacts with a station using a real Hell machine are events to be remembered.

Although CW transmitters can be used to send Feld-Hell, most operation is achieved using SSB transceivers sending keyed audio tones. (The original machines also sent keyed audio tones on telephone lines). The most convenient technique is to use the PC sound card to generate and receive the signals. Software exists for DOS PC and 'Hamcomm' interface, special DSP modules and of course Windows and PC sound card [9]. The most popular software is by Nino Porcino IZ8BLY.

Feld-Hell is spectrally similar to 80 WPM Morse (**Fig 9.7**), and is extremely effective under noisy conditions, copes well with very powerful burst noise and,

Feld-Hell Summary	
Symbol rate	122.5 baud
Typing speed	2.5 CPS (25 WPM) using original font
ITU-R description	350HA1C or 350HJ2C
Bandwidth	350Hz
Modulation	2-ASK (on-off keying, CW)
Average power	22%
Protocol	Quasi-synchronous unconnected scanned image
Character set	User selectable fonts

because of the high peak-to-average power, is effective with simple portable equipment. Feld-Hell is easily generated using a simple microprocessor keyer, so is ideal for beacon transmissions (see Fig 9.6, which is a beacon transmission).

Multi-path performance of Feld-Hell is modest, but reflections can be minimised by receiver-gain adjustment. The effect of multi-path is vertical ghosting on the received text, but the signal is generally still readable. Carrier interference can be a problem. The mode is not seriously affected by Doppler, and receiver tuning and drift are tolerated very well. Feld-Hell has a unique prrt-prrt-prrt sound at letter speed. When the sender stops typing, the transmission stops, and the silence between letters and words is distinctive. The word gaps are clearly seen in the spectrogram (**Fig 9.8**).

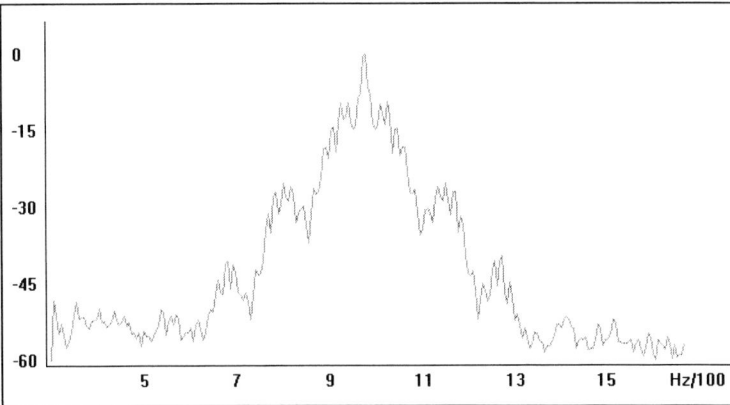

Feld-Hell has never been widely used, but good DX can be worked with patience, and there are hundreds of operators world-wide. At last report Josef OK2WO had worked over 80 countries on Hell! Calling frequencies are around 14.063, 14.080 and 21.063MHz. Hell is more popular on the lower HF bands, but operating frequencies are very regional. Hell has been used for moon-bounce and is especially effective for satellite contacts because of the high tolerance to Doppler and mistuning. There is a 'Hellschreiber Activity Day' every three months and an annual Hell contest, held on 80m and 40m in October (see **Figs 9.9, 9.10**).

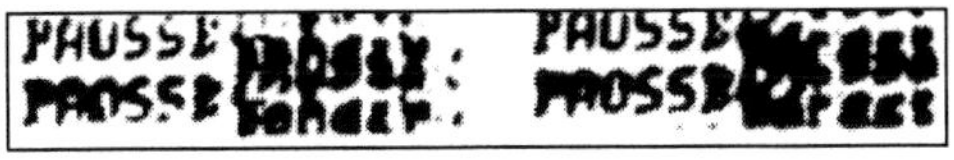

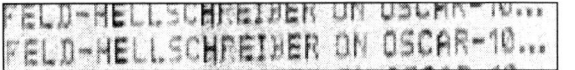

Fig 9.7: The Feld-Hell spectrum

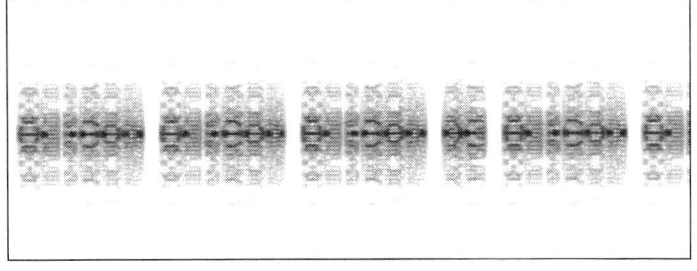

Fig 9.8: The Feld-Hell spectrogram

(left) Fig 9.9: The first Feld-Hell EME reception, PA0SSB 1982 (Picture: Dick Rollema)

(right) Fig 9.10: KD7MW sending Feld-Hell via Oscar 10

Multi-tone Hell

As mentioned above, a seven-tone Hell system was demonstrated in 1937. This system scanned the text in seven horizontal bands, each sent using a different tone frequency. The signal was generated by a series of cams rotating at character speed, and operating switches that keyed seven oscillators spaced across the audio range. More than one oscillator, or indeed all seven, could be on at the same time. The receiver used filters tuned to each of the tones to drive an inked stylus, not dissimilar to a Morse inker, except there were seven styluses. The tones were widely spaced because the oscillators were not phase synchronous, and the filters were tuned circuits with modest performance. (**Fig 9.11** shows MT Hell in use.)

Fig 9.11: Well known digi-mode operator Fred OH/DK4ZC working MT-Hell on 20m

A more modern equivalent was developed by Peter Martinez G3PLX in 1997. The transmission consists of 12 close-spaced tones with raised cosine envelope. All the dots in each column of the character to be sent are generated at the same time, using different tones. With 12 tones the text definition is very good. Reception is via an FFT technique.

There is no particular constraint on how many tones are used, except that they share the transmitter power. Several of the systems available are mutually compatible, despite different dot arrangements and transmission rates. Perhaps the most unusual feature of multi-tone Hell is that it is received using a spectrogram, a frequency-domain display. The text is always correctly aligned, and does not need to be printed twice. The received text moves up or down if the received frequency shifts, and is printed upside down if the incorrect sideband is used!

MT-Hell Summary	
Symbol rate	0–20 baud
Typing speed	2-3 CPS (20–30 WPM) font dependent
ITU-R description	200HF1C
Bandwidth	100–200Hz
Modulation	MFSK (concurrent 9, 10 or 12-FSK or sequential 7-FSK)
Average power	80% (concurrent) 50% (sequential)
Protocol	Non-synchronous unconnected scanned image
Character set	User selectable fonts

All the dots in each column of each character are sent at the same time, and there is no particular timing constraint, except that the successive columns must be sent and received at a speed that gives approximately the right aspect ratio to the characters (related to the number of tones and the tone spacing). Most systems send between two and three characters per second, or about 10 to 20 dot-columns per second or 10–20 baud (20–30 WPM). The speed is dependent on the text size and font used. Most transmissions are designed to be about 200Hz wide. The symbol rate (column rate) and the dot spacing in Hertz are usually the same, for minimum keying interference and maximum throughput for a given bandwidth. The MT-Hell spectrum is shown in **Figs 9.12 and 9.13**.

Fig 9.12: The MT-Hell spectrum

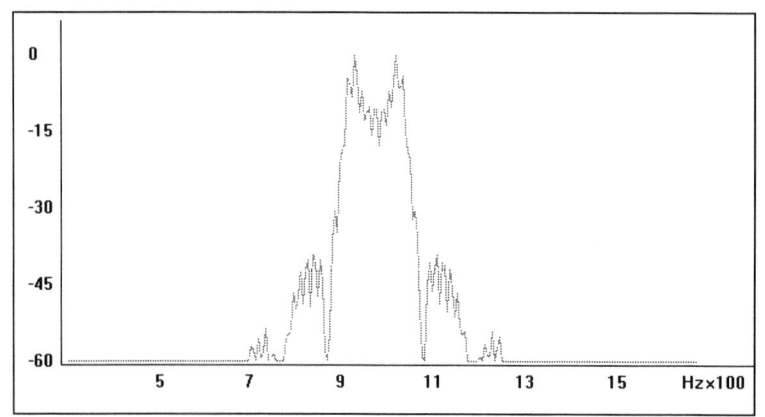

This type of transmission is called Concurrent Multi-tone Hell, and is available in Windows sound-card software by Nino Porcino IZ8BLY, and

DOS sound-card software by Lionel Sear G3PPT. The mode is highly immune to interference, especially carrier interference, but is not very sensitive. At times selective fading is evident as it passes through the signal, but generally does not impede reception. The transmitter must operate very linearly or the transmitted signal becomes very broad, with ghost versions of the text transmitted on either side of the signal. The sound of this mode is a distinctive eee-aww-wee-aww like a squeaking door.

Fig 9.13: MT-Hell can be read in the spectrogram

It is possible to combine sequential dot transmission (as in Feld-Hell) with a multi-tone mode, arriving at Sequential Multi-tone Hell, invented by the author in 1998. The technique scans up the character columns, transmitting black dots as increasing pitch tones. Typically there are seven tones spaced 20Hz and dots are transmitted at 20 baud. White 'dots' consist of silent periods, typically only 30–50% of the duration of black dots, thus increasing the effective speed without increasing the bandwidth. A seven-dot column and up to seven columns per character is typically used, with each character having a maximum of five black dots per column, up to five active columns and two silent columns. A proportional font is used to save transmission time. Through the use of short duration 'white' dots, the proportional font and other dot-skipping techniques, the text-transmission speed is increased significantly. There are typically 11 dot equivalents per character (compare Feld-Hell at 49 dot equivalents). Hence a low baud rate gives a reasonable typing speed.

This mode is little used, but is receiver-compatible with Concurrent MT-Hell (except that the text slopes to the right), and has the major advantages that average power is lower, just one tone is transmitted at a time, and the signal can be generated using FSK rather than DSP techniques. Text quality is limited (**Figs 9.14 and 9.15**), as a low resolution font is needed, but the mode is extremely resistant to carrier interference, multi-path and Doppler.

(left) **Fig 9.14:** Sequential multi-tone Hell has distinctive sloping text

(right) **Fig 9.15:** The author received by EA2BAJ on 20m (note large font)

By sending each column twice, a large easily-read font results. Sequential Multi-tone Hell has also been very effectively used for weak signals on LF using very narrow shifts and very low symbol rates, typically 0.3 baud.

Sequential MT-Hell software is available for DOS and PC Speaker, Hamcomm interface, and parallel-port FSK from the author, and for DOS PC sound card by Lionel Sear Sequential MT-Hell has a most distinctive tweedle-tweedle warbling sound.

PSK Hell

As part of recent research leading to development of new high-performance DX modes, the author and Nino Porcino IZ8BLY explored the possibilities of PSK-Hell, where the black and white dots are transmitted not as amplitude changes, as in Feld-Hell, but as phase changes. In the system developed white dots are

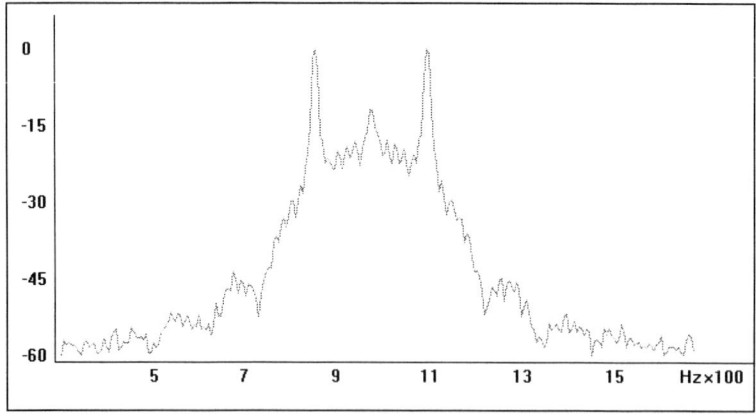

Fig 9.16: The 245 baud PSK-Hell spectrum

signalled as a 180° phase change and black dots as no phase change, thus the mode is differential 2-PSK (DBPSK), like PSK31. Like PSK31, the mode proved to be very sensitive and reasonably immune to interference, but because the receiver operates without synchronism, and phase errors are averaged out by the eye, the mode proved to have very good Doppler-resistant performance as well.

There are two baud rates offered, both speed compatible with Feld-Hell (ie the column rates coincide). The minimum bandwidth version PSK-105 operates at 105 baud, and requires a special low resolution 'differential font'. The 245 baud version PSK-245 uses standard Hell fonts, and even Windows fonts, since the bandwidth is not affected by the same 8ms constraint as Feld-Hell. The signal sounds like Feld-Hell, but with a continuous tone superimposed, or perhaps like PSK31 with erratic warbling – a bee-blee-blee sound. The spectrum of the 245-baud PSK-Hell is illustrated in **Fig 9.16**.

PSK-Hell Summary

Symbol rate	105 or 245 baud
Typing speed	2.5 CPS (25 WPM) font dependent
ITU-R description	210H/490HJ2C or 210H/490HF2C
Bandwidth	210 or 490Hz
Modulation	2-PSK (DPSK)
Average power	80%
Protocol	Non-synchronous unconnected scanned image
Character set	User selectable fonts

The latest version of PSK-Hell has mathematical processing which removes one of the two sidebands normally associated with PSK, and the effect is to increase the average power and impart a characteristic sound rather like FSK. This single-sideband PSK mode has an added benefit of reduced ionospheric distortion on receive (although the mechanism involved is at present not understood). This version (called FM-Hell, eg FM-245) gives best performance and minimum bandwidth. The reception technique, which uses a pair of multipliers and a one-bit delay operating as a phase detector, is identical for PSK-Hell and FM-Hell.

Fig 9.17: Spectrogram of 245 baud PSK-Hell (left) and FM-Hell (right)

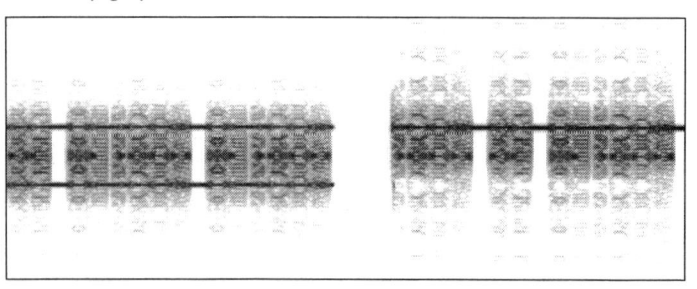

Note that in the spectrogram **Fig 9.17** the FM-Hell signal has the lower carrier missing, and much of the lower-sideband signal has been suppressed.

Fig 9.18: FM-245 PSK-Hell with waves.

Reception of PSK-Hell is similar to Feld-Hell, except that tuning requires a little more precision [10], and there is far less noise. Sensitivity is impressive. The text is invariably very sharp and high in contrast, even when very weak. When experiencing unstable conditions, the received text often exhibits a wave effect, although rarely as severe as the example in **Fig 9.18**. FM-Hell shows the waves more clearly, while PSK-Hell tends to blur under the same conditions. The mechanism of this effect is not yet known.

Note how sharp and clear the text is in Fig 9.18, despite the weak signal (0dB signal to noise), strong Doppler and multi-path on this signal. PSK-Hell is among the best weak-signal long-path DX chat modes, and is certainly the simplest and easiest to use. See the comparisons in Chapter 4 – the performance of PSK-Hell is outstanding.

In the oscilloscope image **Fig 9.19**, there are several white dots followed by a black dot and a further white dot. Notice the change of phase between the white dots, and that the white dots are shaped to zero power at the phase change. Where the black dot is signalled, there is no change in phase, so the signal amplitude is not reduced. This increases the average power. The dot rate amplitude modulation is not used for clock recovery in PSK-Hell, only to control the transmission bandwidth.

Other Hell Modes

Most of the other Hell modes are experimental or historical and now unused. GL-Hell was developed by Rudolf Hell to provide an asynchronous mode of operation. Each character had an initial column of black that started the receiving mechanism, and a white column after the character to stop the mechanism, akin to RTTY. This mode worked well on VHF, although lack of suitable software limited its use. The synchronism is adversely affected by noise on HF. This mode did not leave blank gaps in the tape when the sender stopped transmitting, and did not require the text to be printed twice, as the text was always correctly aligned. GL-Hell is offered by the LA0BX DOS Hell software.

FSK-Hell is just Feld-Hell with an FSK modulator (like RTTY), and is quite effective in reducing noise. The shift used is normally 245Hz, but operation is not common as it is outclassed by PSK-Hell. FSK-Hell is supported by some modern Hell software.

Duplo-Hell is a recent FSK-Hell variation by Nino Porcino IZ8BLY, which uses two concurrent tones to transmit dots in two columns at the same time. The purpose of this is to halve the baud rate for the same text rate, and is quite effective in reducing the effects of multi-path reception. It is not widely used.

The only other Hell modes used to any extent are 1/8 speed Feld-Hell (ie. 15.3125 baud), effective for weak signal beacons and auroral propagation, and Slowfeld, an unusual

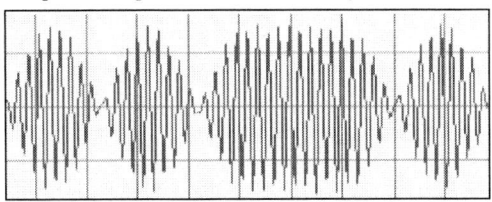

Fig 9.19: Oscilloscope image of PSK-Hell

very slow HF beacon mode by Lionel Sear G3PPT [9]. Slow sequential MT-Hell variants are being developed for LF use.

References

[1] See http://www.qsl.net/zl1bpu/FUZZY/History/History.html
[2] 'Die Entwicklung des Hell-Schreibers', Rudolf Hell, Hell Techniche Mitteilungen no. 1/1940 pp. 2–11.
[3] 'Field remote writer', or field teleprinter.
[4] 'Der Siemens-Hell-Feldschreiber', G. Ege & H. Promnitz, Hell Techniche Mitteilungen no. 1/1940 pp. 11–20.
[5] Stan Cook G5XB, in *Radio Communication*, Hans PA0CX published in *Ham Radio Magazine*, Dick Rollema PA0SE in Reflekties.
[6] 'A seven-frequency radio-printer', 'Electrical Communication', 1937. L. Devaux and F. Smets, Les Laboratoires, *Le Matériel Téléphonique*, Paris.
[7] See Chapter 1.
[8] Using 'raised cosine' shaping.
[9] Software for all Hell modes is at http://home.wanadoo.nl/nl9222/software.htm
[10] If mistuned by 52 or 122Hz, (for PSK-105 and PSK-245 respectively), the text appears white on black.

10

MFSK modes

In this chapter
- How MFSK works
- MFSK16
- MFSK8
- Throb
- FSK441

IF THIS BOOK had been written just two years ago, there would have been no MFSK chapter. While MFSK has been around for many years, Amateur MFSK applications have all arrived within this period. The first was MFSK16, but there are also MFSK8, 'Throb' and FSK441 to consider.

MFSK means Multiple Frequency Shift Keying, and generally means the generation of one tone frequency or radio frequency at a time, although there are concurrent tone systems, such as DTMF, Concurrent MT-Hell and Throb. One of the most well known Concurrent MFSK systems is Dual Tone Multi-Frequency signalling (DTMF) [1], used to send numbers via telephone and to control Amateur repeaters. The DTMF signal consists of eight tones in two groups of four (high and low). Every number is sent as two concurrent tones, one from each group. The character set is limited to 16 combinations, numbers and a few other characters. Hell-family concurrent-tone MFSK modes have already been referred to in Chapter 9 [2].

The history of MFSK modes is nearly as old as FSK, which we call 2-FSK, because there are two frequencies. MFSK modes have typically from four to 32 tones. The reasoning behind MFSK is that with more tones, more data bits can be sent per symbol, and therefore either the possible transmission rate is higher, or more typically, the baud rate is lower for a given transmission rate. This has important advantages, especially in reducing the effect of multi-path reception.

The first sequential-tone systems evolved in the 1950s, in order to solve substantial problems with RTTY links over long distances. The two outstanding examples are Piccolo and Coquelet, both well named because of the sound they made (Coquelet means cockerel). Both existed in several versions.

Piccolo was developed in about 1957 by the British Foreign and Commonwealth Office for the diplomatic service. The system was first publicly demonstrated at an Institute of Electrical Engineers exhibition in London in 1963, where it aroused much interest (see **Fig 10.1**).

Piccolo was a sophisticated electronic system, and the first versions used 33 tones, sending one tone for each character in the ITA-2 alphabet, and the 33rd tone while idle. The system used conventional teleprinter equipment for input and output. By transmitting at ten characters per second the equipment matched speed with 75 baud teleprinters. The signal bandwidth was about 400Hz.

Fig 10.1: Piccolo receivers at Bletchley Park (Photo: Bob King)

Later versions used fewer tones, and sent each character by means of two sequential tones (ie two symbols per ITA2 character). With eight tones sent at 20 baud, a narrower 75 baud ITA-2 system was developed. A lesser-used version with 12 tones sent in pairs was used with the ASCII alphabet. Piccolo was always synchronous, and the symbol clock was transmitted by amplitude modulation of the tones. The signal did indeed sound flute-like.

Piccolo tones were spaced at exactly the baud rate, this being the first such system where this close spacing was possible. To achieve it a special 'integrate-and-dump' detector was developed to receive the signal. This design was highly significant, as it provided at the same time both a highly sensitive and very noise-immune narrow-band MFSK detection system [3]. Piccolo was used for 30 years on long-distance diplomatic teletype circuits, for example Singapore to England; it typically extended reliable operation by several hours each day over an equivalent RTTY system. No error correction was used, but performance was such that none was needed. Although the application was different (fixed circuits), it compared more than favourably with SiTOR, and was less complex. Unfortunately it never achieved popularity in commercial systems. One of the more famous commercial installations was on Cunard's QE2 liner.

Coquelet was developed by the ACEC company in Belgium in the 1950s, and the first machines were electro-mechanical. Tones were generated by mechanically-excited reed filters, and the same reeds were used in the receiver to detect the tones. The Coquelet systems were mostly asynchronous, and the equipment was most often built into a special teleprinter machine.

Coquelet Mk 1 used 13 tones, one for idle, and 12 for signalling, arranged in two high and low tone groups; the former of four tones, and the latter of eight tones. The tones were arranged around a 1052Hz idle tone, spaced by 30Hz per tone. Coquelet operated at 13.33 baud and later at higher speeds. Each character consisted of one low group tone followed by one high group tone. The tone pairs mapped exactly to three and two ITA-2 data bits respectively. This also allowed Coquelet to be generated directly from, or recorded on, conventional ITA-2 punched tape.

Later synchronous Coquelet systems dispensed with the idle tone and synchronism was achieved by filtering the tone pairs.

Because of the electro-mechanical reed filters, and certain keying limitations, Coquelet used a tone spacing three times the baud rate. The signal sounded more like an African Kalimba than a rooster – a fast bong-boing-bong-boing.

Coquelet technology was well documented in an important paper which includes an excellent analysis of why it is advantageous to limit the number of symbols per character to one or two [4]. This paper also includes descriptions and illustrations of the equipment. While not as sophisticated as Piccolo, the Coquelet equipment was relatively simple, reliable and highly portable. It was mainly used by French, Belgian and Algerian military and diplomatic services, generally with encryption equipment also provided by ACEC.

How MFSK Works

FSK, or 2-FSK, sends one bit of data at a time, using two tones. It therefore can be received using a frequency-discriminator circuit, like FM. With more tones, more bits can be sent at a time, for example four bits with 16 tones, but a more sophisticated detector is required.

The detector described by J.D. Ralphs was just such a device. If a high-Q tuned circuit is 'quenched' to stop circulating current, and then audio energy near the tuned frequency is applied, the circuit will start to oscillate, and the amplitude will build up. If the audio energy is at exactly the tuned frequency, the amplitude will build linearly, for some period of time, at a rate dependent on the tuned circuit Q. The Piccolo design used Hartley oscillators adjusted to the point of oscillation, and thus the filters had near infinite Q.

It was found that when the audio applied was off-frequency by a small amount, rather than building up linearly, the amplitude would increase then decrease again, like a modulated sine wave. The time to the first minimum was found to coincide exactly with the reciprocal of the frequency offset. For example, if the tone were off-tune by 10Hz, the first null would be after 100ms. Thus if the tuned circuit response was measured at this time, it would have no response to a tone 10Hz off, or indeed to any tone an exact multiple of 10Hz off frequency, since this tone would have its second, third or fourth etc. null at the same point. This principle was at the heart of the integrate-and-dump detector, and allowed for sensitive detection of tones spaced as close as 10Hz with very little influence from adjacent tones.

Another advantage of the integrate-and-dump system was the sensitivity achieved by detecting signals by comparison, the aim being to locate only the filter which contained the most energy compared with the other filters, rather than having to detect the signal-plus-noise energy compared with prior or subsequent noise in the same channel. Each tone was detected with a very narrow filter, so the noise received was very small and the detector very sensitive. Filters with a few hertz bandwidth or less could be used, since although the signal was up to 400Hz wide, each tone was keyed slowly, at 10 baud. By adding more

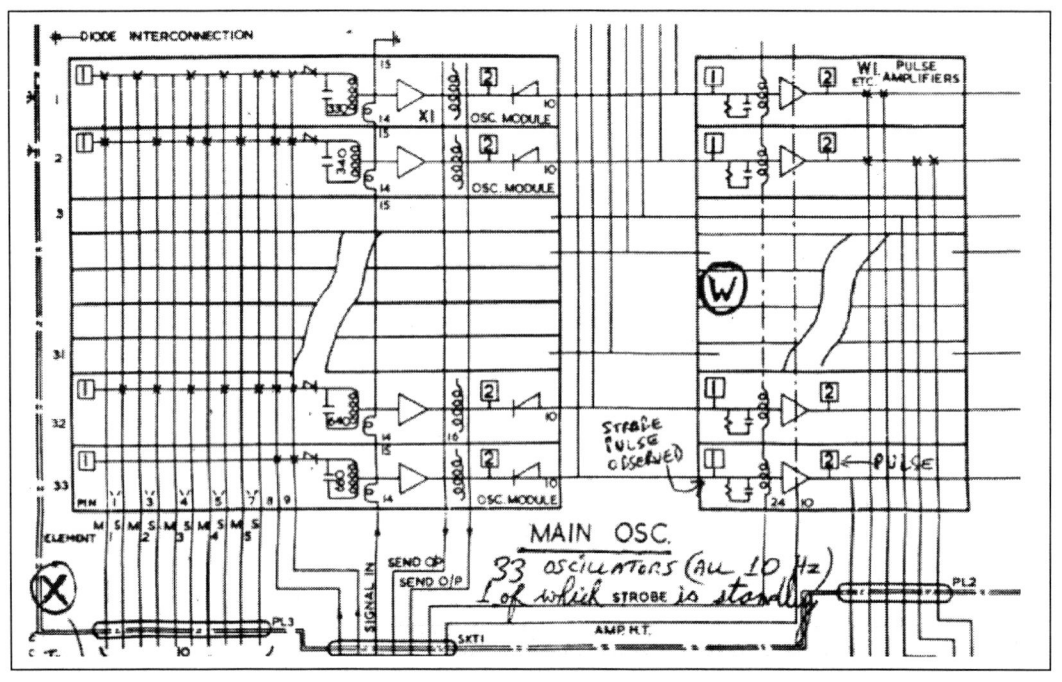

Fig 10.2: Part of the original Piccolo Mk 3 filter-circuit diagram, complete with technician's notes

tones, the overall system-noise performance was not degraded, in fact it was enhanced, when compared with 2-FSK. Part of the original Piccolo Mk 3 filter-circuit diagram is shown in **Fig 10.2**.

The integrate-and-dump detectors were 'quenched' at the start of each symbol period, and the filters 'read' and then 'quenched' again at the start of the next symbol. Of course, for this reason the detector operation had to be synchronised with the transmitted symbols, and therefore required a synchronising system. In the Piccolo system this was achieved by amplitude modulating the transmitted tones slightly. The first half of each tone (or the first tone of a pair) was 10% stronger.

Modern MFSK systems such as MFSK16, MFSK8 or Throb use a DSP software simulation of the integrate-and-dump detector, called the synchronous Fast Fourier Transform (FFT). Repeated groups of samples are taken throughout the symbol period, converted to the frequency domain, and the data summed into 'bins', one per tone, at the end of the symbol period. The result is virtually identical to the Piccolo detector, but is achieved completely mathematically within the computer.

Because of the predictable impulse response of the FFT filters to the square-shaped transmitted tones, it is also possible to extract synchronising information from an MFSK signal using data from the FFT detector. This technique is used in MFSK16 and MFSK8, so no special synchronising modulation is necessary.

MFSK16

MFSK16 is an outstanding example of a modern Amateur chat mode with serious DX performance. It was designed by the author in 1999, and developed for Windows and sound card by Nino Porcino IZ8BLY [5]. This system incorporates the techniques described above and many other modern techniques as well, and is based on a digital bit stream rather than characters. Thus any data can be transmitted. There are 16 tones, spaced at 15.625Hz, and the symbol rate is thus 15.625 baud. This odd number suits the 8kHz sound-card sampling rate. Because each tone carries $\log_2 16$ or four bits of data, the raw data rate is thus 4 x 15.625 = 62.5 bits/sec.

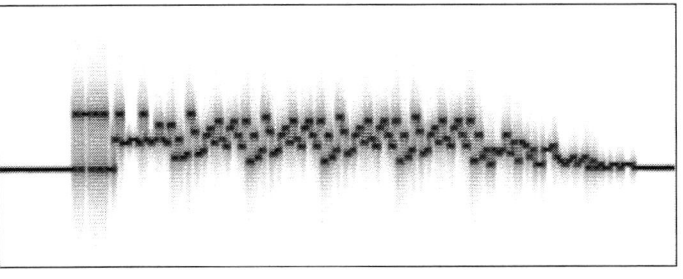

Fig 10.3: The MFSK16 spectrogram

MFSK16 was designed specifically to address the problems which can prevent good-quality long-haul DX conversations, especially long-path DX. It was found during development, and confirmed by many users since, that the MFSK16 mode usually outperformed other modes under weak and unstable long-path conditions, and also gave a handy advantage when operating on low bands with high atmospheric-noise levels and NVIS conditions.

The low symbol rate gives excellent multi-path performance, rejecting delays in excess of 10ms easily. The mode is insensitive to burst noise, and because of the FFT detector performance offers good Doppler tolerance as well.

In order to provide faster text-sending performance, MFSK16 uses a 'Varicode' alphabet, where more frequently-used characters contain fewer data bits and therefore take less time to send. The speed improvement is around 20% compared with ASCII. Another advantage of the Varicode is that very large character sets can be accommodated. MFSK16 has a full extended ASCII set (256 combinations) which could readily be increased.

Full-time error correction is used, in the form of a NASA standard binary convolutional code, a Viterbi decoder (in fact more than one) and an interleaver. One interesting point about the FEC decoder: unlike QPSK31, where the two FEC data bits used to reconstruct each corrected data bit are sent together and identified because they have different phase shifts, in MFSK16 the FEC data bits are sent sequentially. Identifying the bits correctly from the bit order is achieved automatically since the same bits in each group of four are always used (four bits per symbol), ie one of the FEC bits is always the odd bit, the other always the even bit. The interleaver is also synchronised automatically, since the interleaver size is an exact multiple of the number of bits per symbol (x10). The MFSK16 spectrogram and spectrum are shown in **Figs 10.3 and 10.4** respectively.

MFSK16 SUMMARY	
Symbol rate	15.625 baud
Typing speed	4 CPS (40 WPM)
ITU-R Description	316HF1B
Bandwidth	316Hz
Modulation	16-FSK (coherent phase)
Average power	100%
Protocol	Synchronous unconnected bit stream, convolutional FEC
Character set	Extended 8 bit X-ASCII translated to Varicode

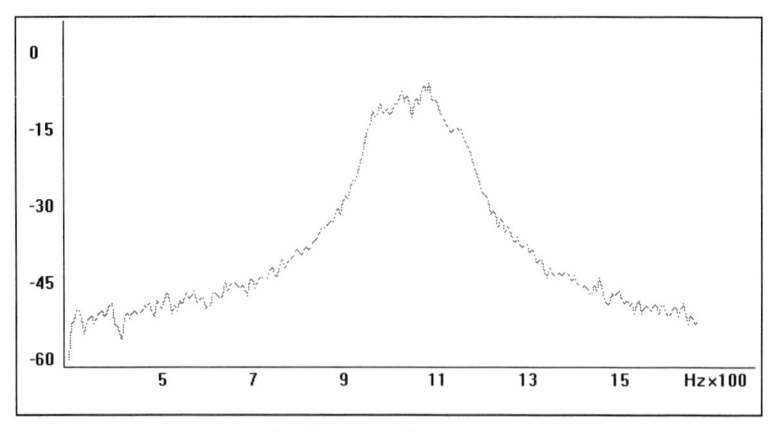

Fig 10.4: The MFSK16 spectrum

MFSK16 was designed for long-path DX conversations, hence the slow symbol rate and narrow tone spacing. A system like this is also very sensitive. The error coding provides essentially error-free communications until the signal disappears into the noise, but reduces the text throughput by a factor of two. The interleaver is used to improve the ECC performance in the presence of burst noise, but results in a six-second delay before the received text is decoded and printed.

The MFSK16 transmission has no amplitude modulation, as the tones change smoothly from one to the next at full power, with no gaps and with no phase change. This minimizes the transmitted bandwidth and eliminates the need for linear amplification, while still allowing the receiver to discover the synchronism very effectively from the FFT detector data.

The FFT response to a step frequency change has a distinctive sinx/x multi-humped bell-shaped envelope. The data from all the FFT 'bins' are summed at several times the symbol rate from the samples taken during each symbol, and the centre of this bell-shape is located by a correlation technique (ie by comparing the data with a known shape). The detector symbol phase is then adjusted slowly to track the symbols. The synchronisation continues to track the symbols as the propagation conditions change.

One important feature of a good MFSK system is the use of a 'guard band' or dead time between symbols. With multi-path reception, the symbols can easily overlap or underlap, causing inter-symbol noise and confusion in the detector, or at least degraded sensitivity. A small amount of the signal at each end of the symbol (typically 5ms) can be left out of the detector calculations, to avoid this overlap. Although sensitivity is potentially reduced, under the poorest conditions reception is dramatically improved. This technique is used in MFSK16.

The MFSK16 sound is distinctive [6], rather like a jumble of musical notes, an orchestra gone mad! MFSK16 needs to be tuned very carefully, and this is done using the software rather than the transceiver.

Operating MFSK16

MFSK16 and MFSK8 modes are, like SSB and RTTY, sideband dependent, in other words, you cannot copy them 'upside down'. Fortunately, most software provides a reversing switch.

By convention, operation on all bands should be with the idle tone as the lowest RF frequency. This will occur when using USB on the transceiver, and the default USB position on the 'Stream' or Hamscope software. MFSK16 and MFSK8 can also be used with a narrow 'CW' filter. However, some transceivers restrict the use of narrow filters for digital modes to LSB, so if it is more convenient to operate the rig on LSB or an FSK setting which generates LSB, simply select the 'LSB' position on the software to invert the tones.

CHAPTER 10: MFSK MODES

MFSK16 and MFSK8 are narrow band digital modes, and should be operated in the same part of the bands as RTTY. However, it is a good idea to keep clear of the traditional calling frequencies, especially 14.080MHz. You will hear the distinctive sound of MFSK16 on either side of 14.080MHz, and also around 21.080MHz and 18.105MHz.

MFSK modes require very precise frequency control, very stable equipment, and very low offset between transmit and receive. As a general guide, drift in excess of 10Hz per over is excessive, and makes communication difficult. This rules out many older VFO transceivers.

Offsets between transmitter and receiver in excess of 5Hz require retuning every over, and beyond 10Hz will make contact very difficult. This problem cannot be corrected using RIT. Requirements for MFSK8 are even tighter.

Partly because of the accurate tuning requirement, there is an 'idle tone' transmitted at the start of each over. Don't be tempted to suppress it, or delay the start of TX, as this tone helps the other person's equipment to stay on frequency and correctly tuned to your frequency. Similarly, don't cut the transmitter prematurely at the end of an over, or your last few words will be lost (remember the six-second delay).

MFSK16 and MFSK8 use very strong forward-error correction (FEC), but at the cost of increased delay between overs. MFSK QSOs are consequently a leisurely activity. Leave plenty of room between overs, in case someone wishes to join in. When tuning in a signal, wait a few seconds to see if correct printing occurs, before trying again. With practice the correct tuning spot can be found on the first or second try. See the following section on tuning MFSK16.

NEVER use RIT on MFSK. Use the software fine tuning, not the VFO on the transceiver. Even if your receiver will tune in 1Hz steps, it is better and more convenient to use the software tuning and AFC.

Tuning MFSK16

The lowest tone is the one to look for. This is the idle tone which is visible at the start of transmission, and from time to time during the over. Observe the picture (**Fig10.5**) which is from the IZ8BLY Stream software waterfall tuning display.

Notice how, toward the left of the figure, the lower thin tuning line is exactly centred on the thick line of the signal, which is the idle carrier (lowest tone). Further to the right, the lower thin tuning line is also exactly centred on some black vertical stripes with fainter grey stripes above and below. It is the idle tone and these lowest black stripes that you tune to, and centre on, not the grey ones above and below. At the same time, the upper thin tuning line is also centred on some of the black data stripes, but this is less important. This picture is shown with the waterfall in zoom x3 mode.

The tuning display is of the point-and-click variety. When the mouse is moved over the display, you will see two black lines that you use to bracket the received signal, moving with the mouse. Move these to the correct spot (lower black line exactly on the lowest tones as described above) and then click with the left

Fig 10.5: The Stream waterfall tuning display (detail)

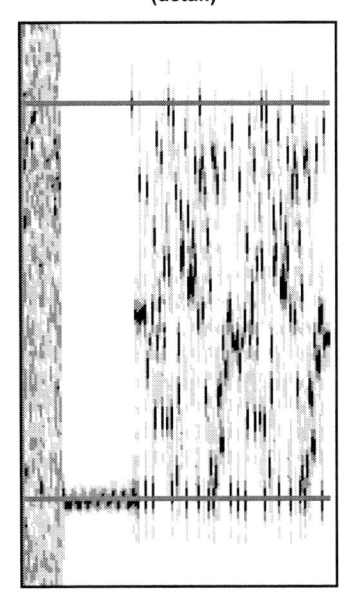

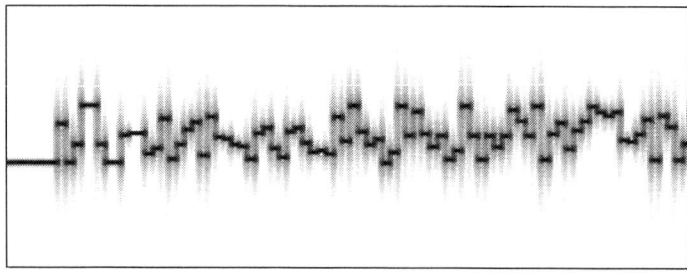

Fig 10.6: The MFSK8 spectrogram

mouse button to fix the tuning. Click so the lower black line is exactly in the centre of the lowest tone, just like the red one in the picture. This will move the red lines so that they coincide with the black. (If the signal is not already correctly tuned, the red lines will be in the same place, but the signal will be somewhere else). Unlike PSK31, Hellschreiber or MT63 waterfalls, the red lines are fixed – it is the waterfall that moves.

Wait for a moment after clicking, and you will see the signal move in from the right to a new position, one hopes correctly tuned. After a few seconds, correct text will be decoded and displayed in the receiving window. You can also change the tuning with the small frequency window below the tuning display, or the up–down arrows associated with it. These are useful for correcting slight drift or offset.

Tuning needs care, but the results are worth the effort, and the technique is easily learned. The only problem is tuning really weak signals. There are signals that cannot be seen or heard, but will still print quite well!

MFSK16 software

The first, and arguably the best, program for MFSK16 and MFSK8 is Stream by Nino Porcino IZ8BLY. This software is something of a development platform, and also offers PSK31 and a number of unique experimental modes such as PSK63F. Hamscope by Glen Hansen KD5HIO, and the MIXW 2.0 multi-mode program by Nick Fedoseev UT2UZ and Denis Nechitailov UU9JDR, also operate MFSK16. All these programs require at least a Pentium processor, sound card and Windows 95. Timo OH2BNS has a version of MFSK16 for Linux.

MFSK8

Designed by the same team as MFSK16, this mode is a little slower, but more sensitive. It is in all other respects similar to MFSK16; the software used and the signal bandwidth are the same. The tuning requirement is a little tighter, and requires extremely careful tuning. MFSK8 uses 32 tones and operates at 7.8125 baud. It has been found to be an effective mode for extending a QSO when band conditions are fading out. The signal sounds like slow electronic music. The

MFSK8 SUMMARY	
Symbol Rate	7.8125 baud
Typing speed	2 CPS (19 WPM)
ITU-R Description	316HF1B
Bandwidth	316Hz
Modulation	32-FSK (coherent phase)
Average Power	100%
Protocol	Synchronous unconnected bit stream, convolutional FEC
Character set	Extended 8 bit X-ASCII translated to Varicode

MFSK8 spectrogram is illustrated in **Fig 10.6**.

Stream by IZ8BLY is the only software for MFSK8.

Other modes related to MFSK16 are likely to be seen in the future. There is some interest in development of a much 'slicker' MFSK mode (offering easy tuning and lighter FEC with no interleaver latency) for contest use. Experiments with 4-FSK at 31.25 baud have been very promising (it certainly offers easier tuning, and it operates at 40 WPM with FEC). A suitable linear FEC coding system has yet to be developed. The research continues.

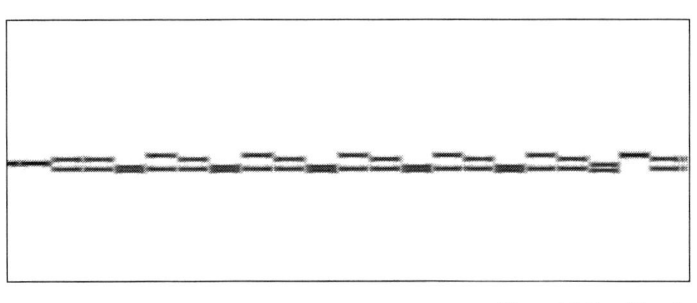

Fig 10.7: Throb spectrogram at 2 baud

Throb

This is an experimental nine-tone system designed by Lionel Sear G3PPT [7]. This mode is unusual in that it sometimes sends one tone and sometimes two at a time. Each tone or tone pair sends one character, ie there is one symbol per character. Throb has a limited character set similar to ITA-2. Because of this, despite a very slow symbol rate, Throb has a remarkable typing speed. There are speed options from one to four baud. The tones are shaped in a sinusoidal manner so that synchronization can be detected. The Throb spectrogram at 2 baud is illustrated in **Fig 10.7**.

The Throb signal is rather narrower than other MFSK modes, only 72 or 144Hz, depending on the speed. The receiver uses FFT tone detection, provides automatic synchronism and AFC. There is no FEC.

MFSK modes are sideband sensitive, in other words must be received the 'right way up', but this problem has been circumvented in Throb by printing the text twice – for normal orientation and for inverted orientation!

Throb is well named – especially when operated at 1 baud. The signal sounds perhaps like a pair of inebriated flute players – whee-whaw-wheee-whaw.

Throb is little used, but operates much as Feld-Hell or RTTY, for DX conversations, and is especially suited to low power, although an SSB transceiver is required. Software for Throb is available from Lionel G3PPT [6], and it is also offered in the MIXW 2.0 multi-mode program by Nick Fedoseev UT2UZ and Denis Nechitailov UU9JDR.

Throb SUMMARY	
Symbol Rate	1, 2 or 4 baud
Typing speed	1, 2 or 4 CPS (10 - 40 WPM)
ITU-R Description	100HF1B or 200 HF1B
Bandwidth	100 or 200Hz
Modulation	9-FSK (sequential and 2-tone concurrent)
Average power	80%
Protocol	Synchronous unconnected characters
Character set	Restricted, similar to ITA-2

FSK441

FSK441 Summary	
Symbol Rate	441 baud
Typing speed	147 CPS (1470 WPM)
ITU-R Description	K21F1B
Bandwidth	2205Hz
Modulation	4-FSK (sequential, three symbols/character)
Average Power	100%
Protocol	Synchronous unconnected characters
Character set	Restricted, PUA-43 (similar to ITA-2)

This wide MFSK mode was designed by Joe Taylor K1JT for meteor-scatter operation, and for other activities where a possible signal path may be present only briefly and with poor frequency stability. Meteor scatter provides communication via reflections from meteorite 'pings', short-lived trails of ionized air in the upper atmosphere. These trails reflect signals throughout the HF and VHF spectrums, but are most widely used on VHF where no viable line-of-sight or ionospheric path exists.

Because the 'pings' can be as short as ten to a few hundred milliseconds in duration, a very high data rate is required to pass as much information as possible. Typically operators set up a schedule where one transmits for the first half of each minute, and the other the second half, and each transmission consists of a small amount of information repeated over and over again.

One factor in operating with reflections from meteorite trails is that they are subject to quite strong Doppler effects, due to the meteorite arrival speed (50–80,000km/hr) and the upper atmosphere wind (up to 1000km/hr). Thus the mode used must be very frequency tolerant, so wide band operation is a necessity. The Doppler shift is of course frequency dependent. The following narrow-band spectrogram (**Fig 10.8**) (called a 'Dopplergram') illustrates 'pings' on HF by displaying their Doppler signature. The line across the centre is the carrier of a broadcast station received fairly weakly 'in skip', and the dark marks with fantastic shapes are Doppler shifted reflections of the carrier caused by meteorite trails. The horizontal scale is in minutes, and the vertical span 20Hz.

Even at 15MHz the Doppler shift of the meteorite reflections is 7–8Hz. At 50MHz it can easily be 25Hz. The faint lines at the bottom of the picture are reflections from aircraft.

Fig 10.8: Dopplergram of meteorite pings received on 15.16MHz

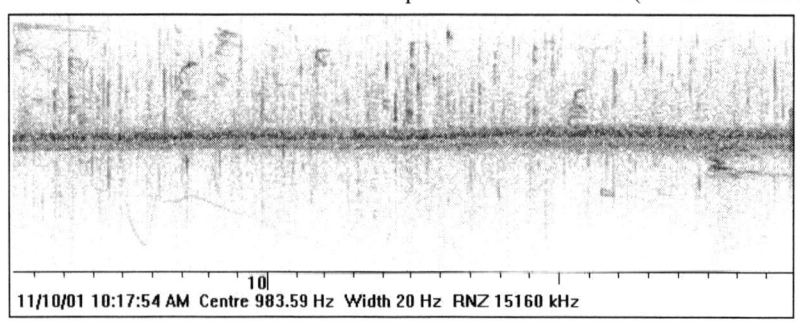

FSK441 is a sequential 4-FSK mode (ie four MFSK tones), operating at 441 baud. In order to manage the Doppler shift, (up to 25Hz on 6m, 75Hz on 2m), the tone frequencies are widely spaced, 882, 1323, 1764 and 2205Hz. Three sequential tones are used to send each character, so the rate

is 441/3 = 147 characters/sec. Of course with only four tones in groups of three, the number of combinations is limited, and like Throb, FSK441 uses a very small character set. The character combinations that would result from all three tones the same are reserved as special 'shorthand message' characters.

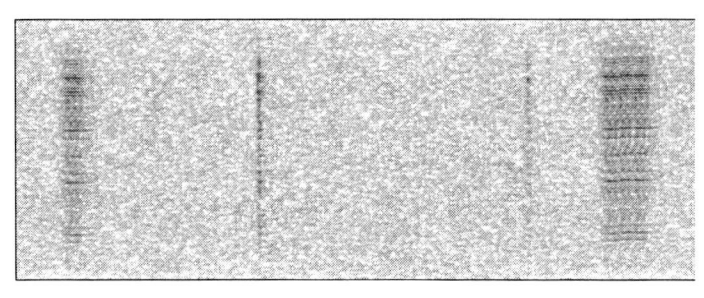

Fig 10.9: Spectrogram of useable FSK441 'pings' in noise, received on 6m

The software for FSK441 is called WSJT and is available from Joe K1JT [8]. Because of the speed at which it operates, it is not practical to decode the received signal in real time. A digital recording is made (a wave file), which is then decoded and displayed. FSK441 has met with considerable success on 6m, and has been tried for moonbounce as well. Because it is an MFSK mode, FSK441 is quite sensitive. A spectrogram of useable FSK441 'pings' in noise, received on 6m is shown in **Fig 10.9**.

References

[1] Developed by Bell Laboratories in the late 1940s.
[2] 'A seven-frequency radio-printer', *Electrical Communication*, 1937. L. Devaux and F. Smets, Les Laboratoires, Le Matériel Téléphonique, Paris.
[3] 'Multi-tone signalling system employing quenched resonators for use on noisy radio-teleprinter circuits', J.D. Ralphs et al, *Proc. IEE* Vol 110. No. 9, September 1963.
[4] 'Les téléimprimeurs, téléchiffreurs et transcodeurs ACEC – système Coquelet', *ACEC Revue* No 3–4, 1970.
[5] See http://www.qsl.net/zl1bpu/MFSK/
[6] Described as 'a strange warble' – see 'Makes Me Wanna Stream', Jack Heller KB7NO, 73 *Amateur Radio Today*, January 2001.
[7] See http://www.lsear.freeserve.co.uk/page3.html
[8] http://pulsar.princeton.edu/~joe/K1JT

11

MT63

In this chapter
- A remarkable mode
- Transmitting MT63
- Receiving MT63
- MT63 software

IF EVER THERE was a remarkable HF mode, this is it. Unlike most of the modes described in this book, which are of modest or even very narrow bandwidth, MT63 is quite the opposite. However, it has some interesting properties that make it very useful under a range of difficult conditions. Perhaps the most unusual feature is the sound of the signal – it sounds like a roaring or rushing noise, often with the atmospheric sound of fading, and it is hard to imagine how with this noise, any data can be transferred. When the signal is strong it sounds like a wall of noise with sharply-tunable edges. This is the juggernaut of DX modes.

MT63 was developed by Pawel Jalocha SP9VRC, in the late 1990s. The author's first QSO was with VK2DSG in November 1998 at a time when the world population of MT63 operators could be counted on one hand. At this time the software only existed for a specialized Motorola DSP unit [1], which operated like a modem.

Almost exactly a year later, SP9VRC released a Linux version, and some months later this was successfully ported to Windows with a graphical interface by Nino Porcino IZ8BLY. The latest version of this Windows software is now the

MT63-1K Summary	
Symbol rate	10 baud
Typing speed	10 CPS (100 WPM)
ITU-R description	1K00J2DEN
Bandwidth	1000Hz
Modulation	64-PSK
Average power	80%
Protocol	Synchronous unconnected FEC with Walsh coding
Character set	ITA-5 ASCII (7 bit)

most popular means of operating this mode, and the mode is more widely used than ever.

The most popular configuration of MT63 is exactly 1kHz wide [2], and has a typing speed of 10 CPS (100 WPM). This mode is commonly known as MT63-1K. There are two other versions, 500Hz wide and 2kHz wide, scaled up and down from MT63-1K. Since the other modes are achieved by exact scaling of all the times and frequencies, they are half and double the speed as well as half and double the bandwidth (halve and double the values in the summary table).

MT63 mode is intended as a chat mode (like RTTY or Hellschreiber), where one station transmits and one or more stations listen. It uses very strong but unusual FEC to provide very good error performance, and is especially good for multi-station QSOs. The signal consists of 64 tones spaced every 15.625Hz across the 1kHz bandwidth, and the tones transmitted and received are in the range from 500 to 1500Hz.

The unique MT63 spectrogram is shown in **Fig 11.1**. It has exactly the same settings and size as those illustrating other modes [3]. The difference in appearance is obvious. Because the tones are so close together, they can't be distinguished in the spectrogram, and as a result a special tuning strategy is also necessary. The signal looks and sounds like noise.

Each of the tones has differential bipolar phase modulation (D-2PSK) and carries 10 bits of information per second, ie the symbol rate of the whole signal is 10 baud. Also, the phase of each tone changes at the same time, as there is no staggering of the clocks. With 64 tones there are 10 x 64 = 640 data bits per second. The data transmitted are 7-bit ASCII, (ITA-5), encoded using a special error correction technique called a Walsh function.

The receiving process is a Fast Fourier Transform (FFT). The FFT can also provide phase information, and this is how MT63 is decoded. By assessing the data in considerably more than 64 'bins' across the signal, a measure of tuning tolerance is achieved.

The Walsh function effectively spreads the seven data bits of each character across 64 bits, and interleaved over 32 symbols, or 3.2 seconds. In other words, the signal is spread in both frequency (so the seven data bits occupy some of the 64 tones), and in time (across some of 32 symbols), and this gives the mode a remarkable ability to reject carrier (frequency) interference and pulse (time) interference.

It also gives the receiver software the ability to decide which tone is which in the received signal. If the signal is mistuned, the expected FFT tone decoders will not give the best result, and a different set of 64 FFT 'bins' including some off to one side or the other will give a better result. The receiver recognises this to be the correct tuning during the initial synchronising period at the start of each over.

Fig 11.1: The unique MT63 spectrogram

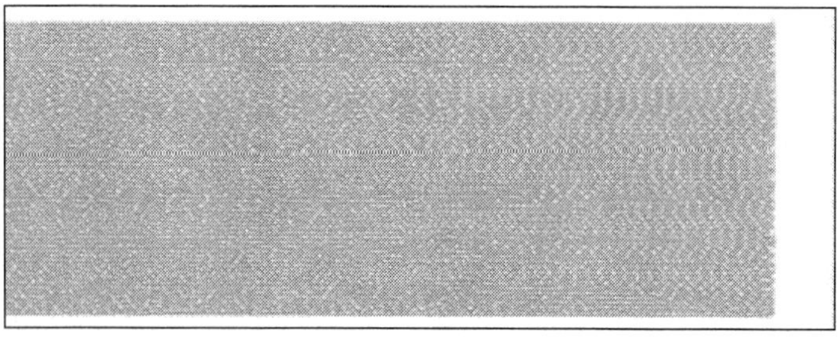

The Walsh function is not at all similar to other FEC systems described here. Perhaps the easiest way to imagine how a Walsh function works is to compare it with a system of posting letters in a postal sorting office that uses 'pigeon holes'. Imagine each character is coded so that it has a maximum distance between its correct 'postal address' value and that of any other postal item [4], in other words if the received letters represent the 'postal address' of a 'pigeon-hole', correctly received postal address letters will place the postal item exactly in the middle of the correct pigeon hole, and the addresses of adjacent holes will be as far away as possible. Now imagine a received character (address of the postal letter) is corrupted, either by a burst of noise or carrier interference. The Walsh function has been designed such that an increasing level of error may cause the 'address' to be inexact, either horizontally or vertically or both, but the letter will still arrive at the correct 'pigeon-hole', since with such strong address coding the errors required to cause a letter to be delivered incorrectly would be huge. Obviously there is a limit to how much noise can be accommodated, but with seven data bits encoded across 64 tones and 32 symbols, this limit is fairly high.

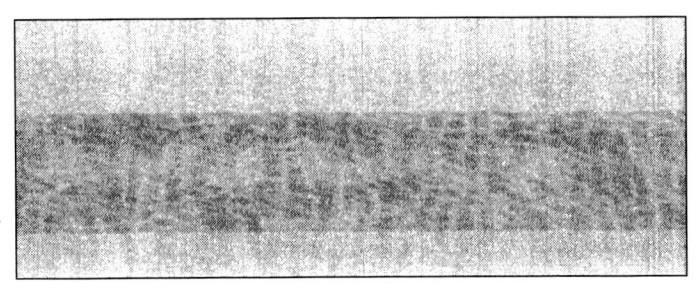

Fig 11.2: Typical MT63-1K signal on 20m – the 'marbling' effect is characteristic

A typical MT63-1K signal on 20m is shown in **Fig 11.2** – the 'marbling' effect is characteristic.

MT63 also comes in different interleaver lengths, 'short', 'normal' and 'long'. Long represents 64 symbols, or 6.4 seconds. This is the most common form found on HF, as it has much better burst-noise performance, at the cost of rather longer delay. Short interleave (16 symbols) is rarely used.

MT63 is good for long-distance QSOs under poor conditions, especially noisy, weak and unstable conditions, and those where there are other stations calling over the top of the signals. The multi-path performance is good, and although the Doppler modulation performance at the low baud rate used should be poor, in practice the performance under these conditions is quite acceptable because of the very strong FEC. The mode operates in the most terrible conditions, and it is not unusual for signals to be read quite adequately when they cannot be recognised – neither seen on the tuning display, nor heard in the speaker!

Transmitting MT63

It is also important to realise that MT63 cannot be identified by ear or eye – while one can assess which width (500Hz, 1000Hz or 2000Hz) is used by observing the tuning display, it is impossible to tell which sideband is in use, or which interleaf is in use, and thus a few simple conventions have been made:

♦ Always use USB to transmit MT63.

♦ Always call using MT63-1K with long interleave (the default), and then change options (if necessary) by negotiation once contact is made.

♦ Since the software tuning is fixed, the transceiver must be carefully netted before calling a station.

In most software versions there is a unique identifier transmitted with every

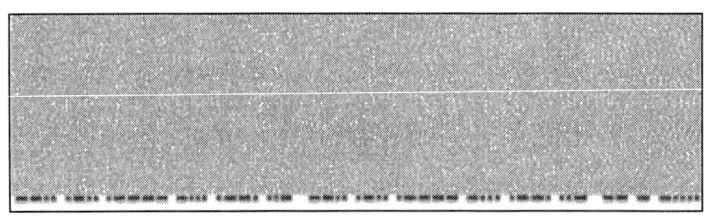

Fig 11.3: CQ transmission with Morse ID on lower edge

CQ call – a Morse ID channel is transmitted along the lower edge of the signal (**Fig 11.3**). This allows the receiving operator to be confident which sideband is in use. The message sends the operator's callsign and the mode ID – 'ZL1BPU ZL1BPU MT63 MODE' is typical.

Since MT63 transmits so many sine waves at the same time, the transmitter must be very linear; most 100W transceivers will need to be operated at around the 25W average level or less to ensure that all the peaks remain linear. Since the receiver is so sensitive, and the mode so immune to interference, this is hardly a problem, and running too much power not only makes the signal too broad, it also distorts the signal and makes it much more difficult to tune and to receive.

When adjusting the transmitter, make sure that the compressor is off and the ALC is not active (not even slightly). With most 100W transceivers 25–30W is about right.

It is easy to fall into the trap of cutting off the transmitter before the signal has finished. Because of the interleaver, there is a considerable delay (typically six seconds) between when the last text is sent as it appears to the transmitting operator, and when the last of the data have finally been transmitted. It is always best to let the software terminate the transmission to ensure that the end of your over and the callsign exchange are not truncated.

MT63 can be described as an 'aggressive' mode. By this is meant that the mode tolerates interference from other transmissions very well, but at the same time can cause considerable interference to those other modes. SSB is perhaps the mode most affected in this way, so it is important to stay well away from established SSB net frequencies. Remember that it might not be your signal that causes trouble to SSB operators you hear, but the station you are working causing interference to SSB net operators you may not hear. When MT63 first became popular it suffered considerable bad press until a suitable home was found. The proper place to operate is wherever wideband digital modes are permitted. If you wish to experiment on these or other frequencies, choose them with care and consultation.

Receiving MT63

The two biggest challenges to receiving MT63 are first of all identifying it, and then secondly tuning it in. There is no substitute for practice, of course. However, a recent operating trend has been to stay on one frequency, and simply wait for other stations to call in (rather like the spider and the fly!), which helps a little. The favourite frequency is with 14.110MHz representing the lower edge of the signal, or 14.1095MHz on the dial of most SSB transceivers in USB mode.

Tuning is performed by observing a fixed-software waterfall display, which in fact displays data from the same FFT that operates the receiving decoder. For example the IZ8BLY MT63 software (see **Fig 11.4**) has blue tuning lines that should be aligned with the edges of the signal. When the mode is changed, the gain of the waterfall display also changes, while the lines stay still. Tuning is achieved by slowly adjusting the receiver tuning until the signal is exactly

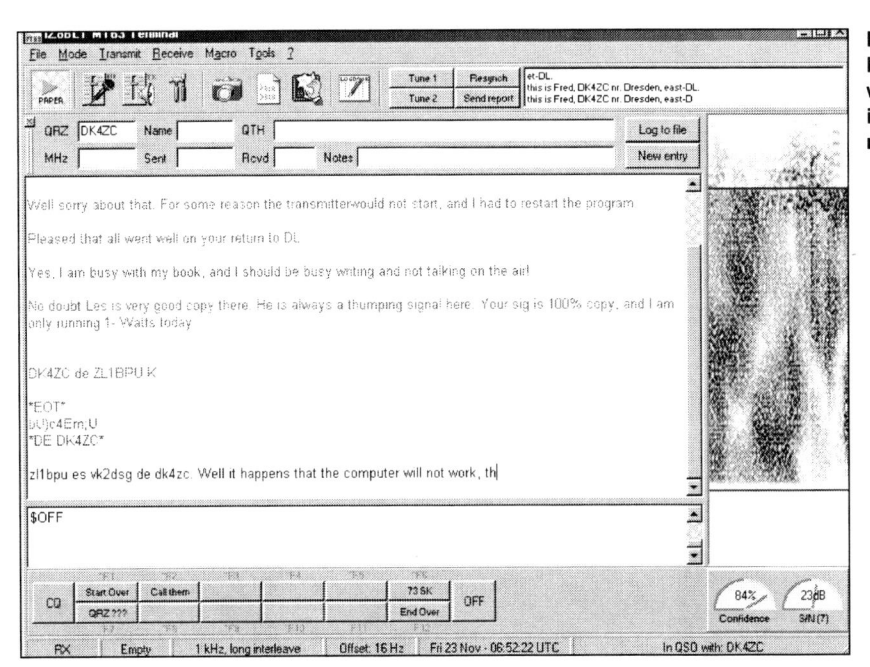

Fig 11.4: The IZ8BLY MT63 software screen (tuning display on right)

between the lines. There is no software tuning, although the software is able to compensate for tuning errors of up to 50Hz.

After a few seconds in the correct tuning position, the software will decide where the correct tones are received, and the 'Confidence' meter below the tuning display will start to rise, indicating that the signal is being decoded correctly. The tuning offset error will have been assessed and displayed at the bottom of the screen. Finally, after as much as 15 seconds, the 'S/N' meter will rise and text will start to print. The very long delay is because of the time it takes the signal to pass through the receiver de-interleaver and decoder. You will also notice that at the end of a transmission, the signal will go away quite some time before the text stops printing.

Once the signal is tuned and printing, do not adjust the tuning again, as even if this is done very slowly or in tiny steps, the software will invariably lose synchronism. The only reason for retuning is if the signal is outside the useable range, and then after correcting the receiver, the 'resync' button should be used to restart the tuning process. Of course all printing will stop until the signal has been acquired again.

When operating MT63-500 (the narrowest mode), the tones are a mere 7.8Hz apart, and the symbol rate is only 5 baud. For this mode the tuning precision required is much higher, and transceiver stability is especially important. The typing speed of this mode is quite good, and normal interleave is more common because of the much longer delays incurred. The wide MT63-2K mode is very tolerant and very fast (embarrassingly fast if you are a slow typist, but great for file transfer). To operate this mode you must have at least a 300Mhz Pentium computer.

Because of the interleaver delays, (typically 12 seconds or more latency between each letter you type and when it is seen on the screen at the other end), MT63 is inevitably a leisurely mode in which to operate. There is no possibility with this mode for lightning QSOs or quick break-in. It is possible to break

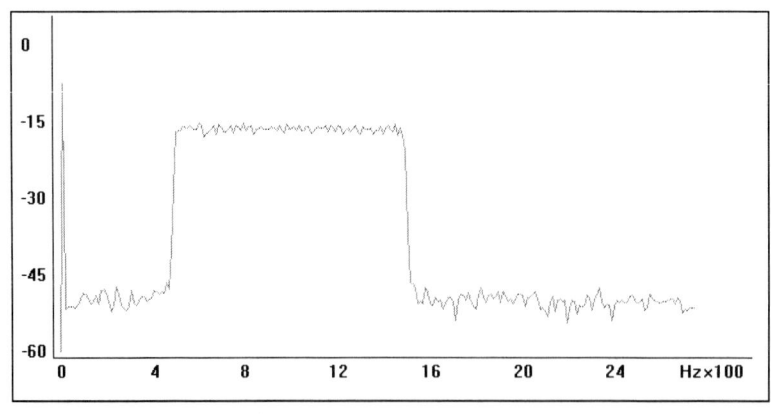

Fig 11.5: The MT63-1K spectrum

MT63 Software

There are several reasonable software options for MT63. The Motorola EVM software is not up-to-date, and the performance is not as good or as easy as the latest Windows software. The Linux software by SP9VRC is a reasonable alternative, but lacks the visual appeal of Windows software. Much the most popular program is IZ8BLY MT63, of course by Nino Porcino IZ8BLY, while MT63 is also now available in the latest MIXW 2.0 multi-mode program by Nick Fedoseev UT2UZ and Denis Nechitailov UU9JDR.

An interesting feature of the IZ8BLY software is the 'secondary channel'. When the transmitting station has temporarily run out of data to transmit (the keyboard buffer is empty), but the transmitter is still active, idle characters must be transmitted to maintain synchronism. Traditionally, the ASCII 0 'null' character is transmitted, but it was discovered by IZ8BLY that in fact any character from ASCII 0 to ASCII 7 would do just as well, without affecting synchronism. This allows three bits of data to be sent for every idle character, so when the transmitting operator is not sending a file or typing at full speed (with a 100 WPM mode, this is most of the time!), this secondary text channel is sending data a few bits at a time. When the main channel is idle, the secondary channel rattles along at about 25 WPM.

Because the secondary data are easily separated out from the main channel text, it is possible to place the output in a little window all by itself [5]. The best use of this data is to send an ID message – typically the operator's callsign, name, location and a simple greeting. The secondary data is a binary bit stream, enjoys the full Walsh FEC coding of the main text channel, and is coded as eight bit data, hence all the extended ASCII characters are available. The main channel text is only seven bit. Software that does not support the secondary channel will still print the main channel normally, and ignore the secondary channel data. The secondary channel data are not visible at the transmitting end, and is preset in a configuration window.

The MT63-1K spectrum is shown in **Fig 11.5**.

References

[1] The Motorola DSP56002EVM, unfortunately no longer available. Some interesting experimental software exists for this unit.
[2] See Fig 11.5.
[3] This isn't true of the other spectrograms in this chapter, Fig 11.2 and Fig 11.3.
[4] In technical terms, this is called the 'Hamming Distance'.
[5] Top right corner of Fig 11.4.

12

PacTOR

In this chapter
* PacTOR ARQ
* PacTOR FEC
* PacTOR II

PACTOR BECAME extremely popular in the early 1990s, both for QSOs and for bulletin-board use. It was developed as a replacement for AmTOR, and while at first it was also used for QSOs, operation now is almost exclusively for bulletin boards. Typically these are used by portable and marine mobile stations wishing to keep in touch with friends via text messages. The mode, like AmTOR, is a bit clumsy for chat operation, but is ideally suited to BBS operation. Like AmTOR, there is an FEC mode which can be used for broadcasts or calling CQ. There is also a newer version called PacTOR II, which offers even better performance. This mode is described separately later in this chapter.

PacTOR ARQ

PacTOR is an FSK mode, operating at 100 or 200 baud and 200Hz shift. It was developed by Hans-Peter Helfert DL6MAA and Ulrich Strate DF4KV in 1991 [1].

PacTOR Summary	
Symbol rate	100 or 200 baud
Typing speed	6.6 CPS (66 WPM)
ITU-R description	400HF1B, 600HF1B
Bandwidth	400 or 600Hz
Modulation	2-FSK
Average power	80% (ARQ), 100% (FEC)
rotocol	Synchronous connected ARQ
	Synchronous unconnected FEC
	Huffman data compression in ASCII mode
Character set	ITA-5 ASCII, 8-bit data

Fig 12.1: The PacTOR spectrum

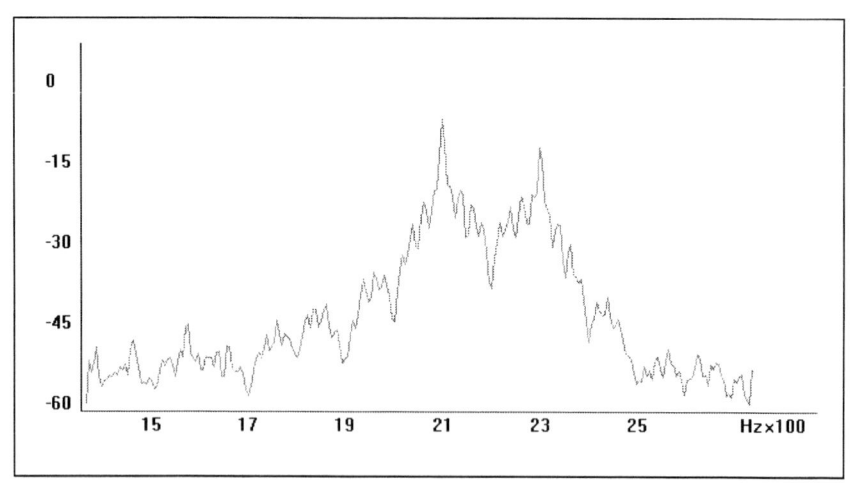

PacTOR is a synchronous connected mode, and operates in a very similar manner to AmTOR. However, the performance is somewhat superior, due to several improvements and innovations. The error correction is much stronger, the frame period is much longer (making long DX and improving data-channel efficiency easier to achieve) and the error correction includes the ability to reconstruct the received information even if the data are never entirely received correctly. Baud-rate changes are automatic, and the receiver filters in the modem can automatically change to provide optimum performance for the receiving speed. Other unusual features are the use of data compression and the use of a signalling technique that makes the transmission independent of the receiver sideband setting.

PacTOR is a commercial mode – there is no free software or hardware method to operate this mode. PacTOR cannot be operated at all with a Windows PC sound card, although there is a software solution for DOS PC and conventional RTTY modem [2]. Most operators use specialised MCUs such as the SCS PTC-II or one of the other devices built under licence from SCS. The PacTOR spectrum is shown in **Fig 12.1**.

Perhaps the most unusual feature of PacTOR is the amazing reconstructive FEC technique. While AmTOR has a relatively primitive parity-type error-detection system (the Moore code), PacTOR uses a very robust error-detection system called a Cyclic Redundancy Check (CRC), similar to that used in AX25 packet radio. This is of course the reason for the name – PacTOR is a packet-radio mode designed for HF.

The CRC is a mathematical algorithm based on a complex polynomial calculation, and generates a 16-bit number that uniquely represents the data sent in each data frame. The CRC is sent (twice) along with the data, and the receiver recalculates the CRC from the received data and compares it with the transmitted CRC value. There is very little chance of the data arriving incorrectly while still agreeing with the CRC, or alternatively, an error being received in the data and in the CRC, and still agreeing. This is certainly not the case with AmTOR.

The special feature of the CRC and the data it represents in the PacTOR design is that, if you were to send one frame of data several times and 'averaged' the data received [3], and also averaged the CRCs received, then as soon as the CRC of the averaged data matched the average of the CRCs received, you could be sure that the averaged data represents exactly what was transmitted. This is important, because it means that the data frame can be received accurately and yet never needs to be received in its entirety!

This technique is called 'memory ARQ', and is used in conjunction with the normal ARQ system. Imagine that a frame is transmitted but is damaged, and the received data do not match the CRC. The receiver will send a 'NAck' (negative acknowledge) asking for a retransmission. However, the received data, while containing some errors, may be substantially correct and are not discarded (as it would be in AmTOR). If the next reception of the same frame is also in error, a NAck should be returned again; but before this is done, the data and checksum are 'averaged' with those from the previous transmission, and the result checked for correct CRC. If the CRC is correct, the data are considered to be accurately received and an Ack (positive acknowledge) is sent, asking for the next frame. If the averaged data are still incorrect, the NAck is sent and the averaging process is continued.

The memory ARQ has the ability to quickly reconstruct slightly-damaged frames without causing a severe reduction in throughput, and is especially effective at countering the effects of burst noise. Without memory ARQ, PacTOR would have to use much smaller frames to ensure sufficient throughput. An interesting feature of the averaging process is that the data from the receiver are 'soft', ie are analogue values direct from the FSK demodulator, rather than digital ones and zeros. This allows the averaging process to accurately add repeated frames of data with great sensitivity, as the process also averages noise to zero. It is not until the averaging has been completed that the data are returned to digital form for CRC calculation and checking.

Each PacTOR frame is 960ms long, and consists of either 96 bits at 100 baud or 192 bits at 200 baud. The frame period is fixed, at 1250ms. The return signal is 12 bits, always at 100 baud, and takes 120ms, received in a 290ms gap, several of which are clearly visible in **Fig 12.2**. The timing is therefore much easier than AmTOR because the receive window is longer, and of course PacTOR does not cause as much wear and tear on the transceiver switching, because the frame period is longer.

The PacTOR Huffman compression is quite similar to Varicode techniques used in other modes. The more frequently-used characters are recoded using fewer bits and the lesser-used characters with more bits. The range is from two to 15 bits. Huffman compression can be turned on and off by the operator, and each frame carries a flag which indicates whether the frame is compressed or not. Huffman compression does not work on data-file transfer, only on ASCII text.

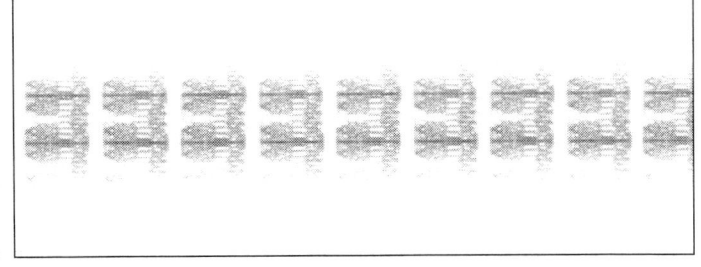

Fig 12.2: The PacTOR ARQ spectrogram

Speed change can be manual or automatic. If several packets are incorrectly acknowledged, the transmitter speed will automatically change down. The receiver can also acknowledge with a speed-change request, forcing the next transmitted

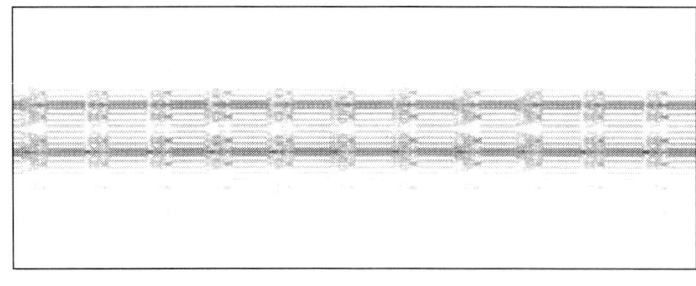

Fig 12.3: PacTOR FEC mode spectrogram

frame to be at the new speed.

The polarity of the transmitted data is reversed for every transmission. The polarity of the frame can be detected from the 'preamble', a unique bit pattern that heads every frame. Apart from the obvious advantage of it not being important to use a particular sideband to receive or transmit PacTOR, this feature also cancels out small receiver offset errors (tuning errors) and, because of the 'soft' detector, allows offset errors and interfering signals to be completely cancelled out during the memory-ARQ averaging process.

PacTOR FEC

The PacTOR FEC mode used for bulletin transmissions or CQ calls is somewhat similar to the equivalent AmTOR mode. The receiving gap is omitted, and the transmitted packets are repeated one or more times. The sender can choose the number of repetitions and also the baud rate.

There is also a 'listen' mode, which is helpful for identifying operating stations, but the performance is not especially good. There is no error-correction capability – only packets with an exact CRC match are printed.

The PacTOR FEC mode spectrogram is shown in **Fig 12.3**.

PacTOR II

PacTOR II is a more recent development that makes greater use of DSP, and yet is designed to be compatible with PacTOR [4]. The calling and linking procedures are identical. Using an extension of the same protocol, PacTOR II soon switches over to more advanced transmission techniques for improved performance. While PacTOR is claimed to be about four times as fast as AmTOR, PacTOR II is claimed to be up to six times faster again. It also has significantly improved sensitivity, since a PSK modulation technique is used.

The PacTOR II signal is only about 500Hz wide, and consists of two phase modulated and raised cosine amplitude modulated carriers spaced 200Hz apart. In addition to a variety of modulation possibilities, PacTOR II adds improved error-correction using convolutional coding.

PacTOR II Summary

Symbol rate	100 or 200 baud
Typing speed	Up to 30 CPS (300 WPM)
ITU-R description	500HG1B(D) or 500HJ2B(D)
Bandwidth	500Hz
Modulation	2-DQPSK
Average power	80% (ARQ), 100% (FEC)
Protocol	Synchronous connected ARQ
	Synchronous unconnected FEC
	Huffman and RLE data compression
Character set	ITA-5 ASCII, 8-bit data

PacTOR II can also use an extended frame time when conditions are good. This extends the performance by reducing the amount of time lost to transmitting control information and waiting for link turn-around. A further enhancement of performance is achieved by using additional data-compression techniques.

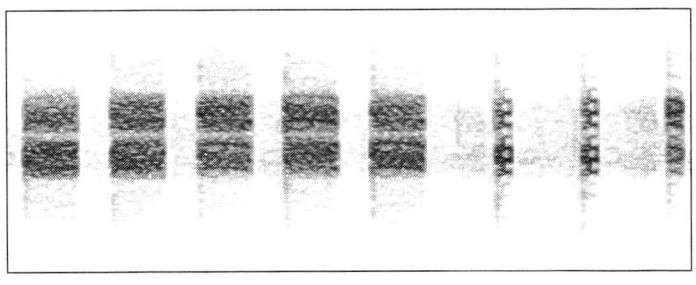

Fig 12.4: Spectrogram of a PacTOR II station in action

The spectrogram in **Fig 12.4** shows off-air reception of a PacTOR II bulletin-board system at the point of changeover. To the left the stronger station is transmitting data (using the normal packet-length option), then it changes to become the receiving station. Note that the longer data packets are sent in PSK, while the short Ack responses are transmitted in FSK.

References

[1] See 'PacTOR, a short system description', The WIA Group, April 1991.
[2] BMK-Multy by Mike Kerry G4BMK.
[3] Let's be intentionally vague about the maths of the averaging process – it isn't important to understanding the technique.
[4] There is a simple description of the development of PacTOR II at http://www.scs-ptc.com/pactor2a.html.

PSK31

In this chapter
- How PSK31 works
- Operating PSK31
- Software for PSK31
- Other PSK modes

PSK31 WAS DEVELOPED in the mid 1990s as a replacement for RTTY. Peter Martinez G3PLX was concerned that the trend in digital modes toward data transfer, message handling and computer linking, made possible by the new ARQ modes, was taking digital mode operation in a different direction to other Amateur modes, which have always been based on two-way conversations [1].

G3PLX thought it was time to bring RTTY into the 21st century by taking advantage of new techniques and new computer processing to make a new mode suited to two-way chat communications. PSK31 was based on earlier experiments with differential PSK by Pawel Jalocha SP9VRC. Peter took the basic idea, added the Varicode alphabet and the raised cosine modulation, and developed high-performance receiving and easy tuning techniques which contribute considerably to the performance and convenience.

Later he added the QPSK mode with a binary convolutional-code error-correction system. During the development, Peter took care to preserve in PSK31 the

PSK31 Summary

Symbol rate	31.25 baud
Typing speed	about 3.5 CPS (35 WPM)
ITU-R description	63HF1B
Bandwidth	62.5Hz
Modulation	Differential 2-PSK (BPSK)
Average power	80%
Protocol	Asynchronous unconnected chat mode
Character set	Varicode, ASCII user interface

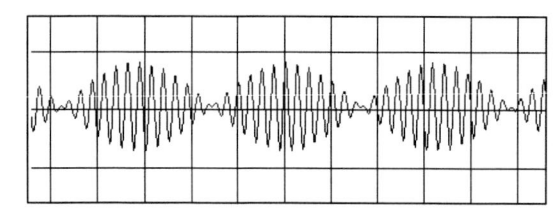

Fig 13.1: Oscilloscope image of PSK31 sending '000'

properties of RTTY which are its best features – quick and easy tuning, and operating simplicity [2].

As a result, PSK31 as we have it today is able to provide better performance with lower power and in a narrower bandwidth that RTTY, and yet is just about as fast and easy to use. The PC sound-card version of the software was released Christmas Day 1999, and since then dozens of other programmers have released software to operate PSK31. The use of PSK31 now far outstrips use of all other digital modes.

How PSK31 Works

The PSK31 modulation concept is really simple [3]. Using a bit stream of data, the carrier phase is reversed for every zero bit, but not when a one bit is transmitted. Because the modulation is 'differential', the absolute phase does not matter, and the data can be recovered by comparing the signal with a delayed version of itself. If the phase is nearly the same as it was 32ms earlier, the datum is a '1', if not it's a '0'.

To ensure that the transmitted bandwidth is kept to a minimum, the transmitted signal strength is at its maximum in the middle of each bit, and reduced to zero at the point where phase reversal may occur, at the start of each new bit. This keeps the switching noise that occurs on phase change from being transmitted. There is of course no reason to reduce the power to zero when the phase will not be changing (ie the next bit is a '1').

To achieve this rounded shape, rather than changing phase by multiplying the carrier with a square wave representing '0' or '1' datum, the carrier modulation uses a special waveform called a 'raised cosine' which smoothly changes from 0 to 1 like a cosine wave. The raised cosine has special minimum bandwidth properties, and is also used by many other modes.

In **Fig 13.1** the sine-wave carrier tone is 100% modulated by the raised cosine wave at the baud rate. Note how the phase of the carrier tone reverses at the point where the envelope amplitude is a minimum. These points are 32ms apart. The horizontal scale markings are 10ms. This is the waveform transmitted when the PSK31 transmitter is idle.

The raised cosine modulation also serves another purpose. The receiving software can easily extract this baud-rate signal, which is then used to control the local-baud-rate clock used to synchronously decode the data.

The PSK31 signal is very narrow. When idle (when '00000...' is transmitted), the signal is like two carriers spaced by the baud rate, ie 31.25Hz apart.

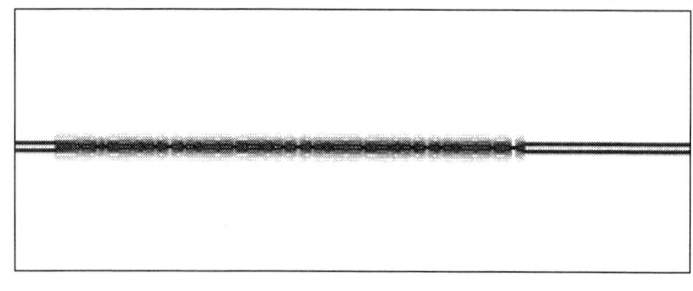

Fig 13.2: The PSK31 spectrogram

When data are being transmitted, the signal consists of a single carrier with 31.25Hz sidebands. It is obvious from **Fig 13.2** how narrow the signal is. At the beginning and end of this trace the transmitter is idling.

CHAPTER 13: PSK31

In **Fig 13.3**, notice how the spectrum of the generated signal is very clean, right down to the floor of the display. It is rather a challenge to actually transmit a signal this clean! The picture consists of two superimposed spectra, the double-humped idle signal and the wider data spectrum with a single hump.

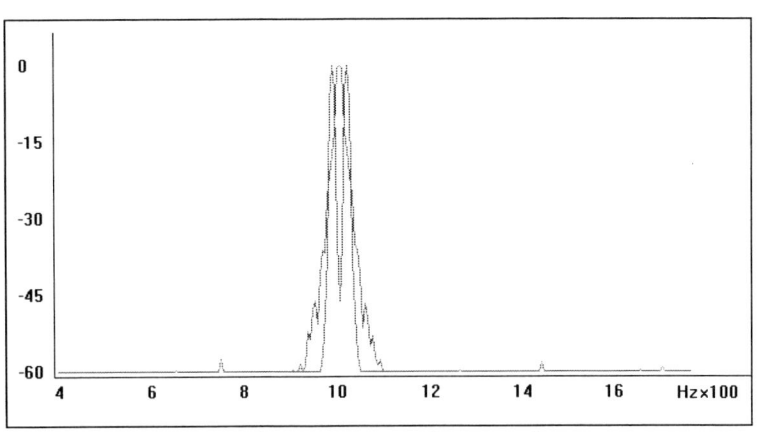

Fig 13.3: PSK31 spectrum, with the idle spectrum superimposed on the data spectrum

The PSK31 data are sent as a bit stream, ie a continuous stream of zeros and ones. You will recall from Chapter 3 that there needs to be some way to determine where one character ends and the next begins. PSK31 uses a Varicode, so the characters are of different lengths. However, they are all separated by the combination '00', which is not allowed to occur anywhere else. Of course when the transmitter is idle, a long series of these '00' groups are transmitted. The smallest useful character is 'space', which is sent as '1' or '100' if the following end of character marker is included. The most frequently used letter 'e' becomes '11', and 't' becomes '101', while the little-used characters (such as 'Q', '111011101') are much longer [4].

Once the received bit stream is decoded, the receiving software simply has to look for an '00' sequence, and decode all the bits from there until the next '00' sequence as a character which it looks up in a table and converts back to ASCII for display.

Operating PSK31

With modern software, PSK31 is so easy to operate that it is typically one of the first digital modes that beginners explore [5]. The software (and there are many versions on offer) is reasonably easy to install, and once the sound-card setup is sorted out, it does not need to be altered again [6].

The first thing to do is become familiar with identifying and tuning PSK31 signals. The signal has a soft warbling sound, like a wobbly whistling note. Most software now has 'point-and-click' operation and very good AFC, so all that is needed is to point at the signal on the waterfall display with the mouse, and click. Text decoding on the screen is almost instant. You will not need to tune the receiver to decode the signal. In fact, you should not adjust the receiver at all – just leave it so that the signal you want is in the receiver passband, and tune in with the software. This is much easier since the software tunes in 1Hz steps, and while a few transceivers tune in 1Hz steps it is by no means easy to tune the signals in this way.

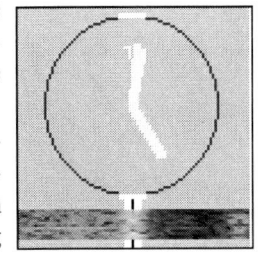

Many programs offer a 'clock' type phase-meter tuning display (see **Fig 13.4**), but generally the point-and-click waterfall type is easier to use. The example given actually includes both. The clock is useful for assessing

Fig 13.4: A typical PSK31 phase meter (PSK31SBW)

117

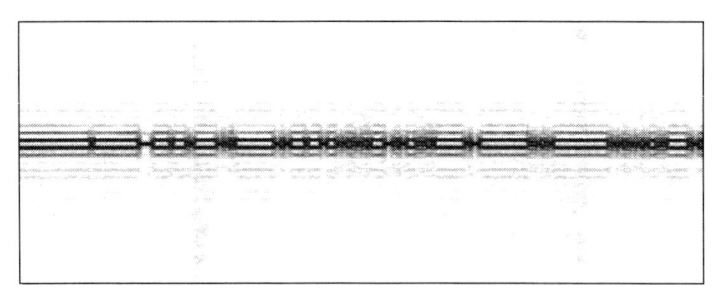

Fig 13.5: An overdriven PSK31 signal (compare with Fig 13.2)

Doppler and drift problems on reception. Older software requires careful attention to the tuning meter, but the recent software is much more forgiving. In QPSK mode the phase meter will show four pointers rather than two.

Some of the low-power transceivers designed especially for PSK31 [7] are crystal controlled, so you have no other way to tune the signals. What happens with these units is that the receiver bandwidth is about 2.4kHz, and you simply choose the signal you want to receive from all the signals available in the passband; there can be 20 or more signals present at once. Digipan software was conceived by Skip Teller KH6TY to provide a specialist tuning display for these transceivers [8]. The display can also indicate the actual band frequency rather than the audio frequency. Some software (Digipan included) also allows you to receive more than one signal at a time. The author has a recorded sample on 20m on a busy evening, and in one minute of recording there are some seven or eight QSOs spread out across the receiver bandwidth. Playing the recording over and over, you can pick out and print one or two different conversations every time. It is hard to imagine doing this with any other mode.

Transmitting PSK31 requires considerable care. The signal consists of closely-spaced sine-wave products, which must be transmitted with great linearity. If the transmitter is not linear, the signal will be audible across a considerable portion of the band, which is unfortunate considering how narrow the signal should be (see **Fig 13.5**). In addition, broader signals are more difficult to tune and usually are not as well received.

In order to transmit linearly, the SSB transmitter should be adjusted to well below the point at which the ALC (transmitter Automatic Level Control) acts. Why is this? Well, the ALC operates as a gain control, in reality a modulator, and the design parameters are chosen to work best with voice SSB. PSK31 has a sinusoidal 31.25Hz envelope which must be reproduced exactly, and this is easily distorted by the ALC modulating the signal. The transmitter gain then changes at the modulation rate, even if the transmitter itself is nominally linear. A 100W transceiver will generally operate very linearly at about 25–30W, and the ALC will not be operating at this level. There is rarely a need to operate more than 25W on PSK31 anyway, no matter what the distance, and there is never a reason to use a power amplifier.

Software for PSK31

Since it is such a popular mode there are many programs available. In fact, it would be no exaggeration to suggest that you could use a different program on each HF and VHF band and still have software to spare [9].

The oldest Windows sound-card software, and still the reference to which others are compared, is PSK31SBW by Peter Martinez G3PLX. This is also the only program which will operate on a Windows 3.1 computer with a 486 processor. Even though there is plenty of sophisticated DSP in this program, the inher-

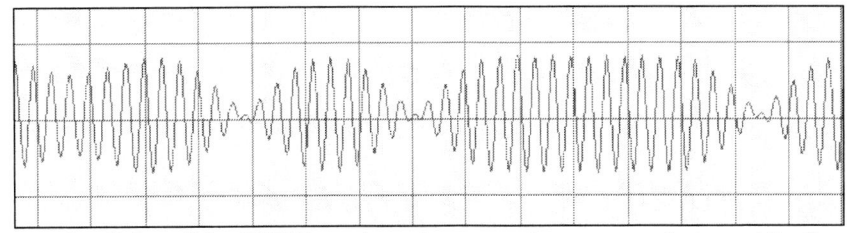

Fig 13.6: The QPSK31 transmission has different amplitude modulation

ent simplicity of the PSK31 mode makes operation on a modest specification computer possible.

Another highly popular program is Digipan, by Skip Teller KH6TY, Nick Fedoseev UT2UZ, and Denis Nechitailov, UU9JDR. This program is particularly easy to use and has excellent tuning, with a large horizontal waterfall display which doubles as the tuning dial when used with a fixed-frequency QRP transceiver.

The versatile multi-mode program MIXW2, once again by Nick Fedoseev UT2UZ, and Denis Nechitailov, UU9JDR, also operates PSK31, as does Nino Porcino IZ8BLY's excellent Stream program. One interesting program, WINPSK by Moe Wheatley AE4JY, has the software written so that the PSK31 part of the program and the user interface are completely separate. Moe has made the PSK31 'engine' available to other developers, along with source code and excellent documentation, and there are now many programs based on this engine, in many languages.

Most of the PSK31 software suggested also operates QPSK mode (with error correction) although this is little used on HF. The reason is that the advantage offered by the error correction is offset by the increased sensitivity to phase errors, especially on longer HF paths and when conditions are poor (remember that QPSK has 90° phase shifts rather than 180°). Thus there is no operational advantage. Another disadvantage is that QPSK31, unlike PSK31, is sideband sensitive. QPSK31 is definitely better where conditions are more stable – LF and the higher bands such as 10m, 6m and of course VHF. **Fig 13.6** shows the QPSK31 signal envelope.

There is software for PSK31 for other computers. There is at least one version for Linux and sound card, at least one for the Mac, and for the oldest version of all, the Motorola DSP56002EVM evaluation board. Some of the more expensive DSP advanced MCUs also offer PSK31. There are even hardware, or partial hardware, solutions, for example a PIC microprocessor-based receiver by Graeme Zimmer VK3GJZ [10].

Other PSK modes

Modes such as MT63 and PacTOR II use similar techniques to PSK31 on a larger scale. There are also other modes that are much closer to PSK31 in design, for example, PSK8, which is exactly the same as PSK31 but scaled down by a factor of four (the baud rate is 7.83125 baud). This mode is of course much slower, and significantly more difficult to tune. It has proved to be very effective on LF where ionospheric phase stability is better and transmitters and receivers less prone to drift.

Nino Porcino IZ8BLY offers several PSK31 derivatives. Perhaps the most popular (although nowhere near as popular as other modes) is PSK63F. This mode uses twice the baud rate of PSK31 (62.5 baud), and uses the extra band-

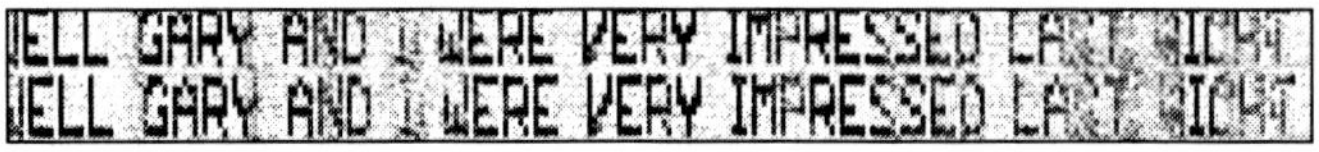

Fig 13.7: The first ever DX PSK-Hell QSO – 15W from ZL1BPU received by IZ8BLY, 21 August 1999

width to code a strong convolutional ECC system. The performance is better on HF than QPSK31 because the higher baud rate is less affected by Doppler modulation, and there are only two phases rather than four. The sensitivity is slightly less than PSK31. Nino's Stream program also offers PSK125F and PSK250F, which are much the same as PSK63F, but faster again. These modes are excellent for file transfer when signals are reasonably strong, but require a powerful computer and stable conditions.

Another variation, FSK31, is offered by UT2UZ and UU9JDR in MIXW. This mode is subtly different, having slightly-different modulation properties, specifically in the phase relationship between the data and subcarrier. The performance under some conditions is better than PSK31, but under most circumstances little difference is observed. To make matters even more confusing, FSK31 and PSK31 look and sound alike. The author is of the opinion that, given the large number of digital modes now available, if a new mode is designed the designer should ensure that the mode should not only offer a benefit, but should be distinctly different from other modes either in sound or spectral appearance, and preferably both.

Bill de Carle VE2IQ offers a wide range of 2-PSK-based modes [11]. These are really for the technical expert, and have been used mostly for weak-signal operation on LF. The modulation, the alphabet and the ECC techniques are all different to most other PSK modes, which are PSK31 derivatives.

In Chapter 9, PSK-Hell was described. While it uses the differential 2-PSK modulation and a raised cosine envelope like PSK31, unlike the other PSK modes, it is not a synchronous data mode. PSK-Hell is non-synchronous, the receiver simply comparing the phase with a delayed version on a continuous basis. The phase differences are portrayed visually, for the eye to assess. The performance is remarkable. **Fig 13.7** shows the first ever DX PSK-Hell QSO – 15W from ZL1BPU received by IZ8BLY, 21 August 1999.

References

[1] 'PSK31: A new radio-teletype mode with a traditional philosophy', paper by Peter Martinez, available at http://det.bi.ehu.es/~jtpjatae/pdf/p31g3plx.pdf

[2] See 'PSK31 – has RTTY's replacement arrived?', Steve Ford WB8IMY, QST May 1999.

[3] See 'PSK31 fundamentals', Peter Martinez G3PLX and Eduardo Jacob EA2BAJ, http://aintel.bi.ehu.es/psk31theory.html

[4] See Appendix B for PSK31 Varicode details.

[5] See 'PSK31 the easy way', Alan Gibbs VK6PG, WIA *Amateur Radio*, March–June 2000.

[6] See Chapter 6 and Appendix D for details.

[7] eg the 'WinWarbler' family by Dave NN1G.

[8] See 'A panoramic transceiving system for PSK31', Skip Teller KH6TY & Dave Benson NN1G, *QST*, June 2000.

[9] See 'PSK31 2000', Steve Ford WB8IMY, *QST* May 2000.

[10] See http://www.users.bigpond.com/gzimmer

[11] See http://cafe.rapidus.net/bill1/bbs.htm

14

RTTY

In this chapter
- History
- Using RTTY
- Hardware for RTTY
- Software for RTTY

RTTY IS ONE of the oldest digital modes, predating all but Morse and Hellschreiber, and has been in use by Radio Amateurs since just after World War Two. RTTY (the name means Radio TeleTYpe) owes its existence largely to the inventions of Jean-Maurice-Émile Baudot, Donald Murray, Frederick Creed, Howard Krum and Edwin Armstrong.

While the history of RTTY is well covered in Chapter 5, we will briefly summarise here the contributions made by these main inventors to what we know as RTTY.

Baudot developed a 5-unit code, patented in 1874, and a manual method of multiplexed signalling, using a five-unit sequential system and a five-key piano-like keyboard. The operator was required to memorise the five-bit code and press the keys in time with the rotation of the multiplexer [1].

In 1903 Donald Murray developed a superior synchronising system for the French telegraph service. Murray won several patents and two important prizes for his work, published several papers, and in 1905 he demonstrated remote type-

RTTY SUMMARY	
Symbol rate	45.45 or 50 baud
Typing speed	6.0 or 6.6 CPS (60 or 66 WPM)
ITU-R description	270HF1B
Bandwidth	270Hz
Modulation	2-FSK
Average power	100%
Protocol	Asynchronous unconnected chat mode
Character set	ITU-R ITA2

Fig 14.1: Mechanical RTTY machines in the shack (photo: Ted Minchin)

setting by telegraph using a synchronized multiplexer, and his own five-bit code, designed to minimise wear of mechanical parts. In time the Murray code became what we now know as the ITA2 alphabet.

Frederick Creed demonstrated a pneumatically-operated telegraph printer to the British Post Office in 1902, and received an order for 12 units. By 1922 Creed was supplying the Post Office with large quantities of start–stop machines for the telegram service.

The US inventor Charles Krum entered the race in 1906, and his son Howard achieved a number of important patents, including one for the first start–stop synchronised teleprinter in 1918.

Until the mid-1950s most teleprinter operation involved direct wire connections, either telephone wires or under-sea cables, because no satisfactory way had been developed to achieve reliable automatic radio transmission and reception of digital traffic. When digital signals were keyed using an on–off technique (the only practical method at the time), it was difficult to distinguish key up and key down because of noise, and even more difficult to follow the strength variation of the signal. The same difficulty still applies today in computer reception of Morse. Mechanical RTTY machines are illustrated in **Fig 14.1**.

Edwin Armstrong patented a system for FM transmission and reception in 1933, and it is from this work that the FSK (Frequency Shift Keyed) technique evolved. FSK allows comparison of the signal plus noise in one channel with noise only in another, overcoming most of the fading and noise problems associated with ASK transmission.

FSK was used to a limited extent during World War Two; in the 1950s, with volume production of teleprinter machines and the reduction of manufacturing cost, teleprinters began to overcome the early lead of Hellschreiber in wireless text transmission, especially for press transmissions. Amateur FSK RTTY as we know it today dates from about 1956. There was some activity in the USA prior to this using on–off keying. Subcarrier (AFSK) techniques date from the 1960s, and the first microprocessor-based decoders from the late 1970s.

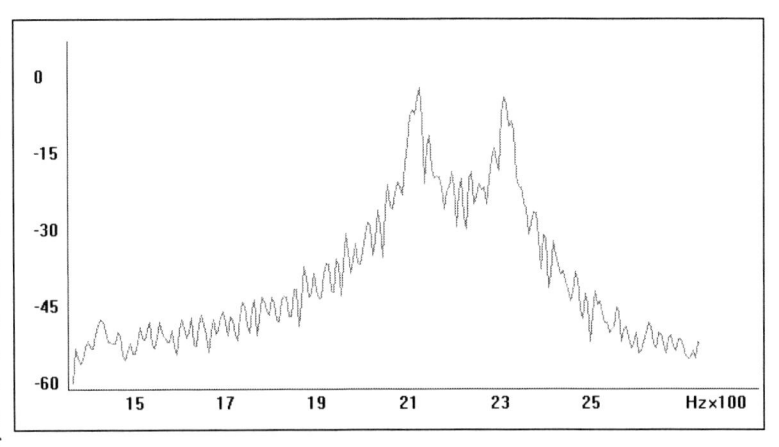

Fig 14.2: The 170Hz shift RTTY FSK spectrum

RTTY Described

RTTY is an asynchronous or stop–start technique, where each character starts and finishes with a special (but unfortunately not unique) bit sequence. In the case of RTTY, each character starts with a '1' bit, which is called the 'start' bit, followed by five data bits, then a '0' which is at least 1.5 bit-times long, called the 'stop' bit [2].

The '0' and '1' levels have traditionally been called 'space' and 'mark' respectively in RTTY parlance. On an FSK radio link the 'mark' and 'space' levels are sent as different audio tones or carrier frequencies, normally 170Hz apart. Each bit is represented as one-tone event, known as a symbol. With FSK modulation and full power applied to the transmitter at all times, the signal spectrum therefore looks like a mountain with two peaks (**Fig 14.2**). RTTY has a distinctive 'beadle-beadle' sound, with pauses on the 'mark' tone between phrases, unless the typist is very skilled or is sending at full speed from a buffer.

Sometimes the transmitting equipment sends occasional non-printing characters while the transmitter is idle. Usually 'LTRS' (letters shift) is sent. There is no purpose served in this except to give the impression that the transmitter is still in use. With older hardware demodulators, this 'diddle' character helped prevent level changes in the threshold corrector that compensated for signal drift.

Fig 14.3: The RTTY spectrogram

The spectrogram in **Fig 14.3** illustrates hand-sent text, and it is clear from the figure when the transmitter is occasionally sitting idle on the 'mark' tone when the transmitting buffer is empty.

Fig 14.4: Many RTTY operators started with the Creed .Model 7. This excellent example is in the BT Museum (Photo: Sam Hallas)

Mechanical teleprinters such the one illustrated in **Fig 14.4** used a shaft with a latching mechanism, driven by a motor via a slipping clutch. When the 'start' bit was received, the magnet caused the latch to be released, enabling one rotation of the shaft, and one complete operation of the mechanism.

During the stop period the magnet was idle and the latch engaged again, stopping the operation. As the shaft rotated, the magnet addressed a selector system, which set up various rods, pins or levers in turn, defining the character to be printed during the next operation. The machines were wonderfully complex.

Modern systems, such as those in computers, frequently use a UART device (Universal Asynchronous Receiver/Transmitter) to achieve the same effect. When the start bit arrives, a clock starts, and it advances a shift register at the centre of each data bit, clocking the data bits into the register. When the stop bit has arrived and has been timed correctly, a message is sent to the computer so it can read the data from the shift register.

Amateur RTTY generally operates at 45.45 or 50 baud, which means that the symbols are 22 or 20ms long respectively. Since one bit of information is sent per symbol, the bit rate is the same, 45.45 or 50bps. The speeds are also traditional – 45.45 baud was used by American teleprinters for many decades. The commercial Telex service operated world wide from the 1960s at 50 baud, and later at 75 baud, until it was overtaken by universal acceptance of FAX systems for business use in the late 1980s. Most remaining data links are now computer-to-computer links or have been replaced by the Internet.

The RTTY character has five data bits, a start bit and 1.5 bit duration stop bit, a total of 7.5 bit times. The typing speed of RTTY at 50 baud is 50 / 7.5 = 6.67 characters per second maximum, which is about 66 WPM. There is no error correction used, so when conditions are poor a certain amount of guess-work is required.

RTTY suffers from two additional problems because of its design. First, reception is badly disrupted for as long as several words if start–stop sync is lost due to noise (remembering that the start–stop sequence is not unique, and several characters may be received incorrectly before the start and stop bits finally coincide at transmitter and receiver). Second, because the ITA2 alphabet is used, there is a great likelihood of garble being printed if a shift character is missed or corrupted. Commercial (and many Amateur) systems get around this by forcing a return to letters shift automatically whenever a space is sent. The space character is present in both shifts. In this way the disruption continues only until the next word space.

Using RTTY

Despite widespread acceptance of newer digital transmission systems, RTTY is still used on a regular basis, mostly for contests and some DX contacts. It is popular for contests because it is easy to use, slick to operate and quick to tune. Slick operation refers to fast and easy switch over from transmit to receive and vice versa, with minimal delay and minimal time to acquire and decode the signal. Many other modes, even new ones, are not as good in this respect.

The most widespread use of RTTY is on the 20m band, where 14.080MHz is the DX calling frequency. Operation is to be found either side of this frequency, and on similar parts of the other DX bands, for example 21.080MHz. RTTY is now rarely used below 20m due to difficulties with multi-path reception, and is rare on VHF where packet radio predominates.

RTTY operation using subcarrier tones has in the past involved two tone standards, the high tones, common in the USA (2125Hz and 2295Hz) and the lower tones (around 1500Hz) more common in Europe. In fact, with PC sound-card software, it really does not matter any more what tones are used. However it is wise to continue to use tones in the upper half of the transmitter filter bandpass, in order to suppress any second harmonic present on the tones. For example, if 1kHz was used and the computer output was clipping a little, the signal would include a 2000Hz component; this would also be radiated, as it is within the 300–2400Hz bandwidth of the transmitter filter.

RTTY operating frequencies are specified as the actual transmitted mark-carrier frequency, which can be measured with a frequency counter. The frequency can be estimated by reading the receiver-frequency display and adding or subtracting the mark audio frequency from the reading, depending on whether USB or LSB is in use.

It is traditional to operate RTTY using an SSB rig in LSB mode. There is now usually no good reason to continue this practice, as the software invariably has the option of inverted operation. Some transceivers, however, will only allow data reception with narrow filters in LSB mode, so the habit may well be perpetuated for some time.

The convention is to operate with the mark (idle) tone higher in frequency, and with most equipment and software this is achieved in LSB with the mark having the lower audio tone, while on USB mark must be the higher tone, achieved by switching the software into reverse. Commercial operation is generally with mark as the lower frequency. Apart from some weather stations and a few press transmissions, most HF RTTY transmissions are now encrypted, and will not be printable.

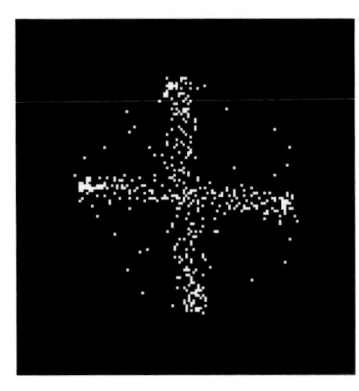

Fig 14.5: The crossed ellipse display

With modern equipment, operating RTTY is simplicity itself. Tune up the transmitter as for CW, set the software in transmit mode, and set the power you wish to use with the microphone gain. When you receive, leave the RIT off and you should be able to print incoming signals and then transmit back on the same frequency without retuning. There are a number of useful tuning displays that make RTTY reception a breeze.

One of the best tuning displays is the crossed ellipse, which has been used for many years. In older equipment the output from each of the two tone filters was used to drive the X and Y deflection of an oscilloscope. **Fig 14.5** shows a crossed-ellipse simulation offered in the MMTTY software. The user simply adjusts the receiver tuning and the software shift and tuning until stable horizontal and vertical ellipses are presented.

If the signal is off-tuned, the deflection is less and the axes are at odd angles, rotating as the signal is tuned. If the shift setting is incorrect, the ellipses will not be at right angles. The crossed ellipse is very effective because it can still be used when signals are very weak, as in the example in Fig 14.5.

Another effective tuning display is the spectrum display, which indicates frequency horizontally and signal power vertically.

This type of display (also from MMTTY) is best when signals are strong, and is used frequently by other modes. The two peaks represent the two tones of the signal, and are correctly tuned when they coincide with the vertical tuning lines. In the case illustrated in **Fig 14.6** the signal is slightly wider than the filters, and has been 'straddle tuned'. The broad 'mountain' under the peaks is not necessarily all signal – much of it is received noise, and illustrates the narrow bandwidth of the receiver and software filters.

A very effective tuning display can be added to older RTTY Terminal Units (TUs) by connecting the signal from each of the discriminator filters to one input of an electronic stereo level meter, especially the vacuum fluorescent or LED type.

The signal is tuned so that the two channels peak at the same time. Good software such as MMTTY has excellent AFC, and it is entertaining to watch the software quickly tune in a signal without user intervention.

Fig 14.6: The MMTTY spectrum display

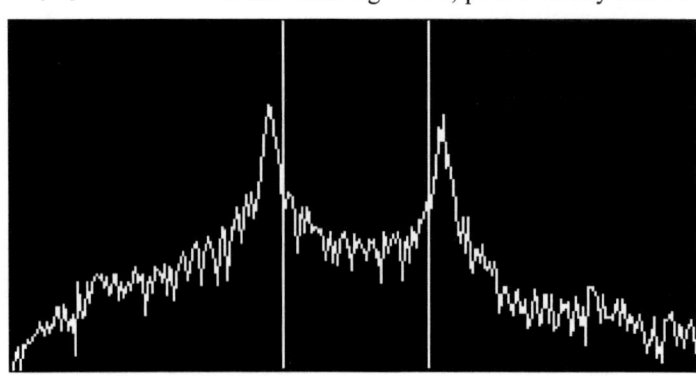

So-called 'CW' filters (400–500Hz bandwidth) can be used very effectively while receiving RTTY, provided they can be centred on the signal when tuned in, or conversely the tuning can be adjusted to suit the filter. Most modern transceivers allow this to take place in LSB mode or in a special data mode.

With modern high performance software, the biggest advantage of using a narrow filter is to prevent receiver desensitisation by adjacent strong signals.

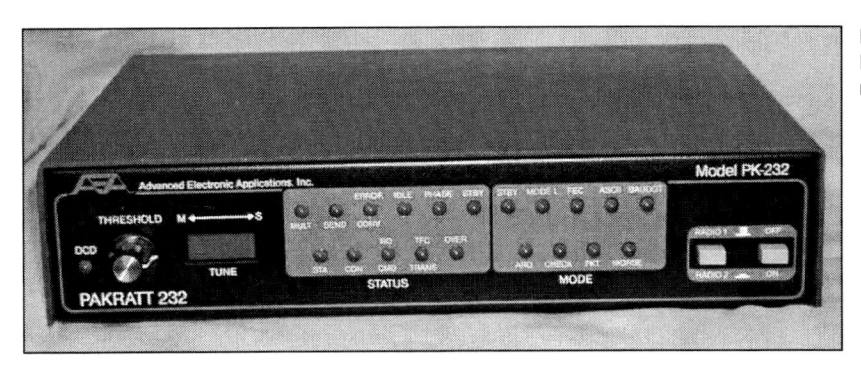

Fig 14.7: The AEA PK-232, a typical multi-mode MCU

The software DSP filters will be sufficient to prevent adjacent signal overload in the computer.

Hardware for RTTY

In the past, 'hardware' might have implied noisy, smelly and very large mechanical teleprinters, and a wall of decoding equipment, but except for a few die-hard 'boat-anchor' enthusiasts, those days are long gone. The hardware of today (or at least the recent past) is the TU, or the Multi-mode Control Unit (MCU), connected to a computer. These units receive and transmit the signals using the subcarrier method, provide some means of tuning in the received signals, and, in the case of the MCU, also convert the digital data into a form more suited to the computer.

The TU (for example the BARTG [3] Easyterm and the Maplin TU1000) contains only the FSK modulation and demodulation hardware, and perhaps a tuning meter, but no 'intelligence'. These units are limited to a very few modes that use FSK, typically RTTY and AmTOR, and perhaps also FAX. The TU needs to be operated with special RTTY software on the computer, set to receive 7.5 bit data at 45.45 or 50 baud. Ideally the filters in the TU need to be adjusted when the baud rate is changed.

The MCU can be a simple unit (like the early Kantronics UTU) operating only RTTY and AmTOR, one of the many packet-radio MCUs, like the AEA PK-232 illustrated in **Fig 14.7**, or very sophisticated truly multi-mode controllers such as the SCS PTC-II, which use special DSP processors and operate many different modes.

During the early 1990s a popular way to operate RTTY and some other modes was through the use of a simple hardware interface widely known as the Hamcomm interface. This unit consisted of a simple comparator limiting circuit for receive, which sent a square wave into the computer via a serial port, and used a simple low-pass filter and PTT circuit for transmit. The unit was very cheap to make, and there was a wide range of software for it. Perhaps the best known was Hamcomm, which provided RTTY mode and a number of useful tools, including a spectral display for tuning and signal analysis. There were also FAX programs such as SCS HF-FAX, JV-FAX and popular programs for SSTV, Pasokon TV Lite and EZSSTV. Unfortunately the performance achieved on RTTY using this simple technique was rarely better than modest, mostly because of the limited filtering possible.

The use of this comparator interface is now largely limited to tuning displays and FAX software.

Fig 14.8: MMTTY copying 75 baud 850 shift commercial traffic

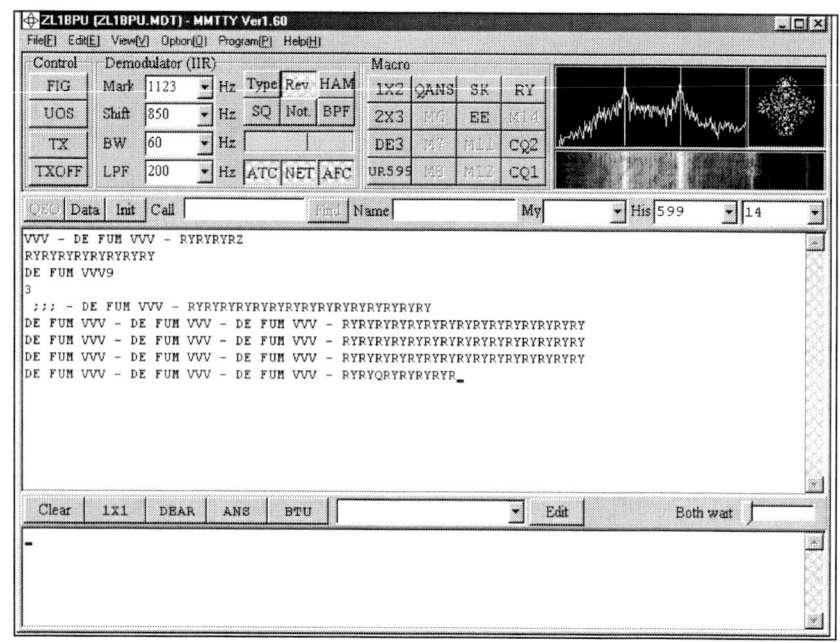

Software for RTTY

Without doubt the best approach for RTTY now is to use the Windows PC with sound card, and there is little argument that MMTTY (**Fig 14.8**) by Mako JE3HHT is the most popular software in use. The performance is very good, the software is easy to use, and there are different settings for typical Amateur use and for receiving commercial RTTY. RTTY is available in a number of other programs, including several multi-mode programs. There is also good RTTY software for DOS and sound card by Rob ZL2AKM.

References

[1] See Fig 1.1, which shows French Baudot multiplex operators.
[2] See detailed description in Appendix A, and especially Fig A.8.
[3] British Amateur Radio Teletype Group.

15

Image modes

In this chapter
- SSTV
  - How SSTV works
  - Operating SSTV
- Facsimile
- Satellite images
- New image modes

AMATEURS HAVE ENJOYED SSTV since the 1950s, so it is natural that this is what comes to mind when image transmission is mentioned. However, Amateurs have dabbled with other image modes for almost as long. FAX has been used in a small way over the years, and it is quite popular for Amateurs to receive commercial FAX transmissions on HF (most of these are now weather maps). Satellite image reception is an especially enjoyable challenge followed by a small band of enthusiasts [1]. The other two image modes explored by Amateurs are of course the multi-frame modes fast-scan TV (normal TV) [2] and various low-resolution narrow-band TV modes. These two are outside the scope of this book.

SSTV

The name of course means Slow Scan TeleVision. Early SSTV was generated using similar techniques to conventional (fast) TV, for example images were generated using flying-spot scanners and vidicon tubes. Receivers used long-persistence cathode-ray tubes, and long-exposure camera shots were the only way to make a permanent record of reception. The early days were definitely the sphere of the dedicated home constructor [3].

The images transmitted in those days really were scaled-down fast TV, complete with repetitive frame images. Despite the technical similarity, with image transmission times of about 16–20 seconds, moving pictures were, however, not possible. The author can recall SSTV in the 1960s where some stations were equipped with home-brew cathode-ray-tube receivers, and transmitted their images from audio tape! The images would be made by recording signals from other stations, or by recording signals generated and personalised by a friend with SSTV signal-generation equipment.

DIGITAL MODES FOR ALL OCCASIONS

Fig 15.1: SSTV reception can be near perfect on VHF

The second phase began with the introduction of frame-store techniques. Homebrew, kitset and commercial equipment (such as the Robot 70 receiver and solid-state sampling cameras) were introduced during the 1970s. This period introduced SSTV to a wider group of Amateurs, and it was this period that marked SSTV as a pursuit for the well-to-do, since the commercial equipment (and even the kitsets) were quite expensive. Colour SSTV arrived during this period, and in 1984 the popular Robot 1200 colour receiver (which was a single-frame system) finally brought repetitive-frame SSTV to an end.

The 1990s were the period of computer involvement in SSTV, with special hardware for computer image display and image generation. Systems were still commercial in nature, and generally expensive.

Around 1995 simple SSTV for the mass Amateur market finally arrived, first with the DOS and Hamcomm interface software Pasokon TV Lite (and the shareware version EZSSTV) and later MSCAN. In 1997 MSCAN SSTV repeaters appeared and finally in 1999 excellent PC sound-card software for SSTV became available. Most of these programs are commercial products, and moderately expensive, but MMSSTV, released in 2000, offers an easy-to-use, high-performance free system for colour SSTV in most of the modern modes (see **Fig.15.1**).

How SSTV Works

Unlike conventional TV, modern SSTV only transmits one image frame. Thus it is a mode for transmitting photographs and other still images. Because of the bandwidth requirements (SSTV has always been a system with voice bandwidth signals), the picture is transmitted slowly, and colour information is generally transmitted sequentially to keep bandwidth down.

The SSTV image (which in modern systems is invariably a computer file image) is scanned sequentially from top left to top right, then each successive line down the picture. Generally there are 250 to 256 picture lines, but since there are so many different modes there are many variations [4]. Several lines are transmitted per second, commencing with a line-synchronising pulse, followed by the image information, generally three sequential scans, either red–green–blue, but other encoding techniques are also found.

SSTV Summary	
Symbol rate	Mode dependent, 200 - 500 baud (pixels/sec)
ITU-R description	1K80F1C or 1K80J3C
Bandwidth	1800Hz
Modulation	Analogue FSK
Average power	100%

The data is transmitted as a frequency-modulated subcarrier. The sync pulse is 1200Hz, and the data occupies 1500Hz (black) to 2100Hz (white). With video sideband data, the signal fits well in an SSB transceiver pass-band (2.1kHz). The FM technique was developed very early in the history of SSTV, and has been used almost exclusively ever since, because it is reasonably sensitive and very effective at cancelling noise. In recent years there have been experimental transmissions using digital techniques.

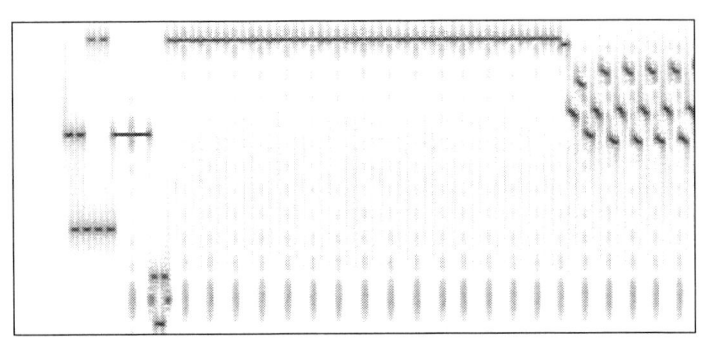

Fig 15.2: The SSTV spectrogram

Fig 15.2 shows the spectrogram of the start of an SSTV picture. The first few marks are the Vertical Interval Signal (VIS) which identifies the mode (see later), followed by a few lines of white (black line near the top) with clear sync pulses showing (at the bottom). At the right side the colour content is visible, while the sync pulses continue.

At the SSTV receiver, the audio is limited to remove noise, then demodulated using an FM discriminator technique. The sync is extracted with a simple comparator and low-pass filter. The video information is then sampled in time with the sync and stored as image data. The PC and sound-card systems replicate this process exactly using DSP techniques, and have the advantages of considerable flexibility, very high performance and great operating convenience. For example, the filters in software such as MMSSTV surpass any that could be realised in hardware, excellent visual tuning aids are available, and the software allows images to be edited and made ready for transmission, while images are being received.

The SSTV spectrum is shown in **Fig 15.3**. The broad 1200Hz sync and three colour peaks are obvious. At the start of each picture a slow sequence of tones is sent. This sequence, called the VIS is unique to each mode, and consists of an MFSK identification for the mode. This allows the receiving software to select the correct mode automatically.

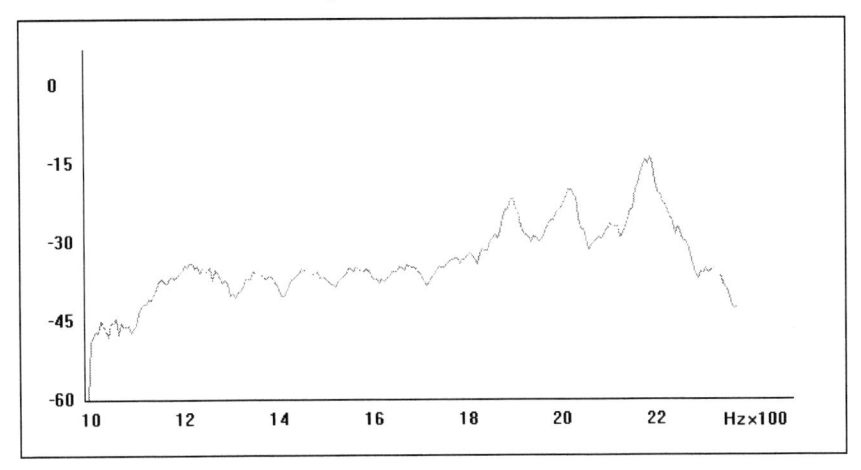

Fig 15.3: The SSTV spectrum. The broad 1200Hz sync and three colour peaks are obvious

131

There are many different SSTV modes, some faster than others, some higher resolution than others, and some better for different conditions. By far the most common modes are SCOTTY 1 and MARTIN 1. Both are reasonably fast general-purpose modes. There is a comprehensive list of modes and specifications on the Internet [5, 6].

Operating SSTV

Essentially SSTV is an adjunct to SSB voice communications. Most QSOs take the form of swapping stories and pictures, and it is rare for a QSO to be solely carried out by picture exchange alone, although the author has experienced several where language difficulties exist. In this case the ability to add name, location and signal-report information to the images transmitted is essential.

The interface between the computer and the transceiver was described in Chapter 6. Because swapping from voice to image is an essential part of an SSTV QSO, it is very helpful to be able to switch from microphone to digital interface instantly. There are a number of simple designs that allow easy switching between microphone and computer output. A very simple solution is to add a small relay and microphone socket to the PTT interface as described in Chapter 6 and illustrated in Fig 6.9.

Tuning up a transmitter for SSTV is easy. Leave the transceiver microphone gain at the normal SSB setting, so there are no adjustments required to use the microphone. Simply adjust the sound-card audio level and gain adjustment as in Fig.6.9 for the required power output. It is best to leave the sound-card setting where you have it for other digital modes; about half way up is a good compromise. If you build the suggested interface, mount it in a small box with the gain-control adjustment accessible from outside the box.

SSTV is usually operated with an SSB transceiver; however, on VHF, SSTV pictures are frequently exchanged using SSB or FM transceivers, and sometimes even via VHF repeaters. This is a very convenient technique, and of course means there is no tuning to worry about. Results can be excellent (see Fig 15.1 and **Fig 15.4**).

Fig 15.4: SSTV Image received from MIR, January 1999 (Picture: David Blackett)

SSTV has also been transmitted very effectively from space, from imaging satellites, from MIR and the International Space Station, as the example received by Dave ZL1AD attests. We can be assured of further exciting developments in Amateur space imaging.

There are a number of SSTV repeaters around the world, most of them using FM on 10m, VHF or UHF, but a few operate on HF SSB as well [7]. SSTV repeaters are simplex repeaters, or 'parrots', and are useful for checking out your own setup extending your operating range. To trigger the repeater, send a short 1750Hz tone, then switch off and wait. If the repeater is ready to record, it will send 'K' in Morse. Then send your picture, preferably in one of the more common modes. Once the picture has completed, switch off, and you will then hear (see?) the repeater repeat your complete picture.

To use an SSTV repeater on HF the transmitter tuning must be very precise. Adjust a few hertz at a time and send the 1750Hz tone repeatedly until the 'K' is heard. Bear in mind that although the repeater will effectively double your range, the useful throughput is only half that of a normal QSO. Also, since the repeaters send their own identification pictures from time to time, and send them in the mode last used, it is not a good idea to use the repeater in a very slow mode, and not then follow it with a normal (faster) mode picture, as the repeater will then send its own ID pictures very slowly until it is changed again. The transmitter will be transmitting almost constantly, and may well overheat.

Fig 15.5: The VK6ET repeater operates on 15m

To receive SSTV, carefully adjust the tuning until the pictures appear to have the correct colours, and the pictures start automatically. The better software has an FFT tuning display, which makes identification of the sync pulse easier. Simply tune until the sync pulse is at 1200Hz. Generally the received picture quality will be reasonable if the tuning is within 50Hz. Some software is very clever, and will identify and start receiving a picture even when tuned in part of the way through the picture.

SSTV picture reception is often poor, even if SSB reception is good. Noise and fading (causing the affects shown in **Fig 15.5**) are a problem, but multi-path reception can also make reception unreadable, as the signal timing can vary by a significant amount of the time taken to send one line, making the pictures appear jagged.

Facsimile

Very few Amateurs use facsimile (FAX), and even fewer have ever transmitted it. Amateurs wishing to send high-resolution images now usually send them as digital files, either by packet or other digital-mode file transfer, or via the Internet!

Most Amateur FAX transmission and reception over the years has used various commercial FAX machines, many of them adapted to suit HF operation, by

HF FAX Summary	
Symbol rate	Typically 1000 pixels/sec
ITU-R description	2K080F1C or 2K00J3C
Bandwidth	2000Hz
Modulation	Analogue FSK
Speed	60, 90, 120, 180 LPM (120 most common)
IOC	288, 576
Average power	100%

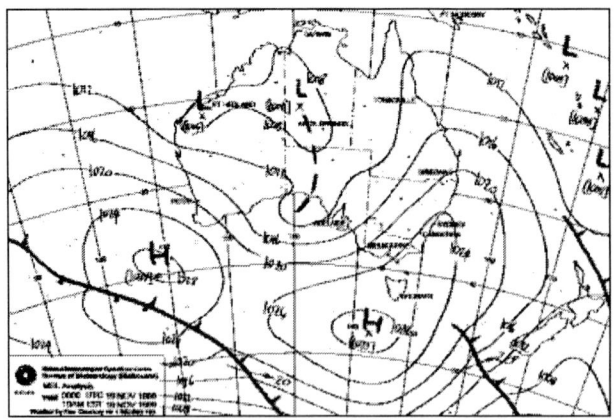

Fig 15.6: A typical HF FAX weather map

changing the scanning speed and the modulation technique. FAX transmission standards are measured by two main parameters – line rate in lines per minute (LPM) and the width-to-height ratio, which is called the Index of Cooperation (IOC). Almost all of the older land-line FAX machines transmitted AM subcarrier images, at line speeds of 240 LPM or greater, and an IOC of 288. Amateur transmissions have usually been made using an approximation of the HF weather facsimile standard, 120 LPM and IOC 576 or 288, and with FSK modulation. The FSK modulation standard is similar to SSTV, although there is no sync pulse sent at the start of each line. Sync is achieved by sending a black bar with a white gap for about 30 lines at the start of each picture. Pictures are normally scanned left to right, top to bottom.

The resolution of FAX images as transmitted on HF can be high – equivalent to well over 1000 picture elements (pixels) per line. Compare this with SSTV, which typically has 100 to 250 pixels per line. However the pictures are very slow to transmit, and times as long as 10 to 20 minutes are typical. Some FAX equipment is capable of very good grey-scale transmission and reproduction, but this is generally not the emphasis of this equipment, intended mostly for reception of black-and-white text and weather maps (see **Fig 15.6**).

The HF FAX spectrogram is illustrated in **Fig 15.7**. The white carrier is obvious as a black line near the top.

Modern Amateur FAX transmissions (and there are very few) are typically computer generated and received. One simple but still useful FAX transmitting program is the older DOS program PC HFFAX 6.0 by SSC, which uses a Hamcomm interface. The most popular software for receiving FAX is WXSAT by Christian Bock, and is receive only.

Satellite Images

Satellite imagery is a pursuit followed by a small band of Amateur enthusiasts. It is a rewarding but time-consuming hobby, but need not be expensive. The cheapest setup would include a simple omnidirectional antenna for VHF, a VHF FM receiver with 30–50 kHz bandwidth [8], and a computer with suitable software. Twenty years ago there would have been many months of construction involved to set up a receiving system, with mechanical receivers, image format converters and a digital frame store system.

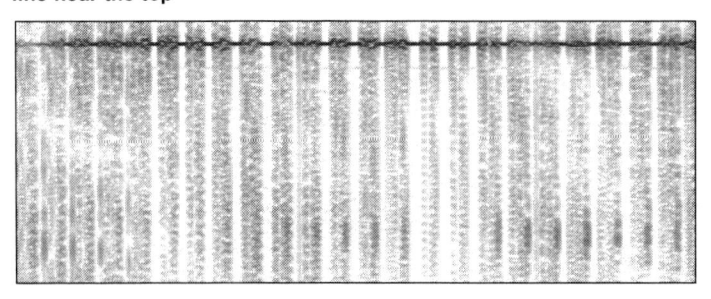

All this has changed, first with special analogue-to-digital conversion cards inside a PC running special software,

Fig 15.7: The HF FAX spectrogram. The white carrier is obvious as a black line near the top

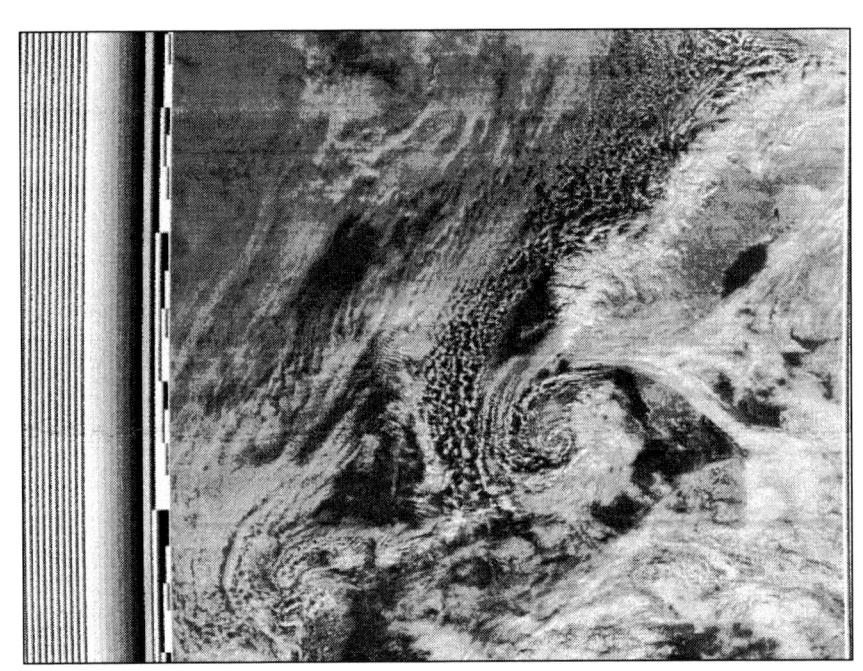

Fig 15.8: Northern Europe from Russian satellite Meteor 3-4

then with simple interfaces such as the SSC unit that plugged into the PC serial port [9]

The most recent software gives high performance reception with a PC sound card. There are many commercial programs available with very high specifications. Most include the ability to add false colour, land outlines and reference grids.

Despite the modern simplicity of equipment, it takes considerable skill and months of patient experience to produce high-quality satellite images. It is a worthwhile challenge, however, as the results can be stunning.

Fig 15.8 shows the resolution possible with very simple equipment. This image was received by the author in April 1995 on 137.4MHz from a polar-orbiting satellite about 1000km above the earth. It shows most of northern Europe, with the UK in the centre. There is a depression over Denmark which brought a cold northerly air flow and snow showers. The mountains and fjords of Norway are very clear. Iceland is at the top left. The bars down the left side are for synchronisation and calibration purposes, and illustrate the grey-scale resolution possible with this setup. The software was PC Goes/Wefax, which has lower resolution than some software, but has excellent noise rejection. Because this software does not sample the pixels synchronously with the sub-carrier, there is a slight curve in the picture, which represents the changing time of arrival of the signal as the satellite passed overhead. The receiver, demodulator and antenna used were homebrew.

There are several high-resolution formats, including digital transmissions, from both polar-orbiting and geostationary weather satellites, but the low-resolution pictures like the one above are the easiest to receive. They use a 2400Hz AM subcarrier system and an FM transmitter. The signals can be quite strong and so are easily received, although a good circularly-polarised antenna is required to prevent fades from ruining the images. The signal bandwidth is about 30kHz, and the subcarrier is synchronous with the image, giving 2400 pixels per line.

In the polar-orbiting satellites, which orbit at an altitude of 800–1000km, and pass over from horizon to horizon in about 15–20 minutes, the horizontal scan is performed optically in the satellite, but the transmission is continuous and the vertical scan is achieved purely by the orbital motion of the satellite over the earth. Transmissions are all around 137MHz. Some satellites transmit a single image at a time, while others transmit two images side by side. Visible, infra-red and water-vapour wavelength images are transmitted.

The geostationary satellites are of course much higher in altitude, and transmit a series of fixed-size square images from a single viewpoint high above the earth. They transmit at 240 LPM on about 1700MHz. To receive this frequency a receiving converter is usually mounted on the antenna, and feeds a 137MHz receiver.

New Image Modes

Over the next few years new image-transmission techniques will emerge. There have already been some excellent demonstrations of high-definition digital SSTV, with perfect reception over very long distances [10]. Before long these techniques will become available to the average Amateur for use with the PC sound card.

Other Amateurs have been developing analogue and digital techniques for narrow-band TV, low-resolution TV modes capable of handling fixed or slowly moving low-resolution images. There are many different formats being explored from simple spinning-disc Baird-type televisors to DSP systems transmitting a complete frame at a time [11]. Without doubt the best digital imaging systems will use advanced modems, good ECC techniques and data compression.

References

[1] Many are members of the 'Remote Imaging Group' (RIG), which publishes an excellent magazine. See www.rig.org.uk/
[2] See http://www.batc.org.uk/index.htm
[3] The history of SSTV is covered at the web site: http://www.darc.de/distrikte/g/T_ATV/sstv-history.htm
[4] Most of the modes are listed in Appendix B.
[5] See http://www.ultranet.com/~sstv/modes.html
[6] See http://www.mindspring.com/~sstv/barberpresentation/proposal_1.htm
[7] There is a list of repeaters at http://www.mscan.com/repeater.htm
[8] It is an annoying fact that most Amateur FM receivers, and most scanners, are not suitable.
[9] 'PC Goes/Wefax', by Software Systems Consulting, 1990.
[10] See http://www.mindspring.com/~sstv/hdsstv/sstv-cmp2.html
[11] See http://users.pandora.be/ON1AIJ/

16

Advanced digital modes

In this chapter
- PC-ALE
- STANAG 4285, 4529
- Q15X25
- Digital voice communications
- Ionospheric sounding

IN A BOOK of this type, this chapter is the one most likely to date. Advanced modes are a popular subject for of ongoing research and development. Various groups are exploring ECC techniques with better efficiency and higher performance, compression techniques to allow higher data rates, new modulation systems, clever ionospheric measurement, sounding and path assessment systems, adaptive modes, new ways to implement existing commercial and military standard modes, and a host of other ideas. In this chapter we will look at a few of the recent systems that have been made available to interested experimenters. Not all of these systems will work on a PC with a sound card at present; many require considerably more sophisticated hardware. Few are regularly used on HF, but all have been tested there.

PC-ALE

Developed as a military system, ALE (Automatic Link Establishment) is more a tool than a mode as such. It started life as US Federal Standard FS-1045A, and became MIL-STD 188-141A. Although not a new development, it is widely used by military networks of all types to track and log communications links.

At the heart of ALE is a simple but reasonably robust MFSK transmission protocol. Eight tones are used, 750Hz to 2500Hz, in 250Hz steps. Each tone carries three data bits, and the tones are

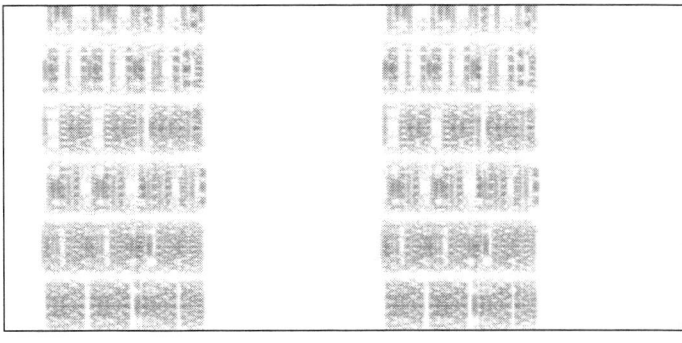

Fig 16.1: The ALE signal is much wider than the spectrogram!

DIGITAL MODES FOR ALL OCCASIONS

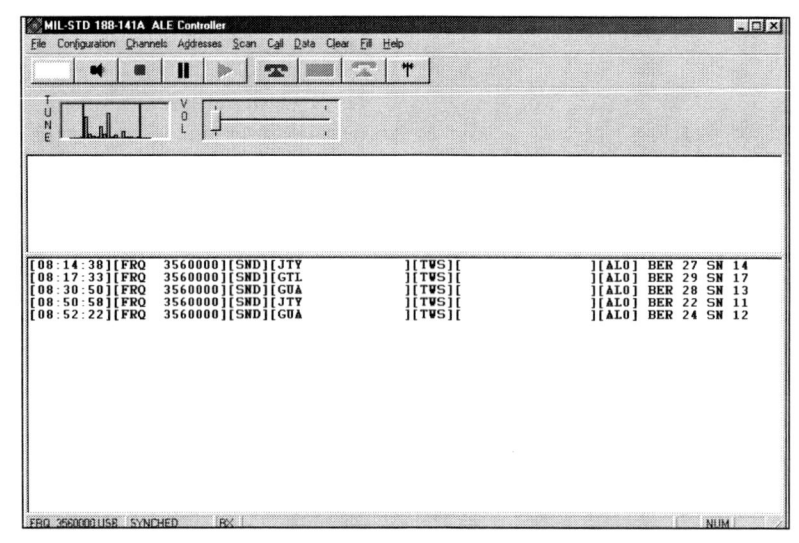

Fig 16.2: G4GUO's PC-ALE software in action

transmitted sequentially at 125 baud. The signal is like a chortling sound, perhaps like a group of excited turkeys in full voice. If you listen just below 20m, you are likely to strike these sounds. They typically only transmit for a few seconds. **Fig 16.1** shows a partial spectogram of the ALE signal, which is much wider than the figure indicates.

If that was all there was to this system, all would be well, but many layers of complexity are added which are beyond the scope of this book. It is enough to say that a protocol has been developed that uses repeated messages with FEC to allow the stations in a network to assess the signal quality and error rate for all the other stations in the network. The network can operate on a single frequency, or, with scanning receivers and remote frequency control, can operate on many frequencies.

A number of specialised messages have been devised, which allow beaconing, polling, routine status transmission, selective paging calls and simple text messages to be transmitted to any or all stations in the network. The power of the system resides in the ability to automatically build a current database of the network, providing virtually instant access to any station.

With ALE it is possible to hold a keyboard-to-keyboard conversation, using an ARQ protocol similar to PacTOR or packet radio, but the protocol is very clumsy and despite the high baud rate, the data rate is low. Most ALE systems are used purely for status reporting, beaconing and selective calling. Once contact is established, the operators move to another mode (which may be digital, but frequently voice), and perhaps another frequency. G4GUO's PC-ALE software in action can be seen in **Fig 16.2**.

ALE is most effective with mobile communications networks, such as aircraft and other military transport systems. There is a very interesting and fully-func-

PC-ALE Summary	
Symbol rate	125 baud
ITU-R description	2K25HF1B
Bandwidth	2250Hz
Modulation	8-FSK
Average power	100%
Protocol	Synchronous connected ARQ
	Synchronous unconnected FEC
	MIL STD 188-141A
Character set	ITA2

tional PC-ALE implementation for PC sound card by Charles Brain G4GUO. It includes scanning facilities for a number of modern transceivers. This is very clever software, looking for an application [1]! Perhaps someone will adapt the excellent 375bps modulation scheme to provide a more appropriate TCP/IP or AX25 Amateur ARQ mode.

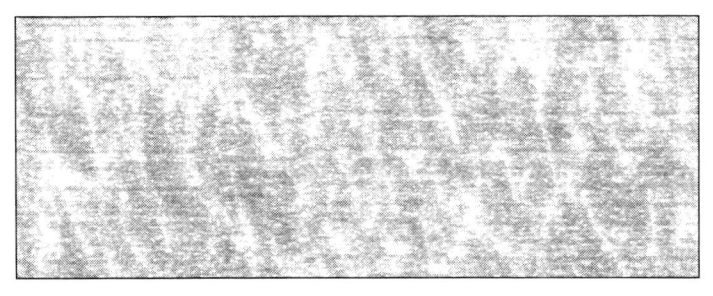

Fig 16.3: Part of the STANAG 4285 spectrogram. Note the marbling effect

STANAG 4285, 4529

These are military standard modem specifications, intended for very high performance on HF. STANAG 4285 is quite widely used by commercial and military networks (but not on the ham bands!), and it sounds rather like very wide MT63, a roaring noise. Quite frequently STANAG 4285 is identifiable as a raised noise level and 'clouds' or marbling on a spectrogram (see **Fig 16.3**), rather than any actual audible signal. The signal is 3kHz wide.

STANAG 4529 is a cut-down version of 4285, occupying half the bandwidth and offering half the data rate. Experience has shown that the performance on real transmission paths is worse than half the speed of the 4285, however.

These modes are highly adaptive and use an advanced FEC protocol. They are also suited to ARQ techniques. Modulation schemes, coding schemes, interleavers and baud rates can be changed automatically to provide rates from 75bps to as high as 3600bps. Ranging and channel-measurement signals are embedded in the system, allowing the link to be assessed and equalisation to be continually adapted during communications.

At present there is no Windows PC sound-card software implementing these modems, but Charles G4GUO has an experimental Linux system operating STANAG 4285 with a TCP/IP protocol stack, simulating a high performance packet radio network on HF [2]. Although adaptations have been made to the protocol to suit TCP/IP, the waveform is identical to the standard.

A related mode, MIL-STD 188-110A, has been explored for Amateur use by Bob McGwier N4HY [3].

Q15X25

Developed by Pawel Jalocha SP9VRC, Q15X25 (sometimes known as NEWQPSK) owes a lot in its concept to MT63. It is an ARQ system that operates at 83.33 baud using differential QPSK, and 15 parallel tones spaced 125Hz apart. The raw data rate achieved is 2500bps. Because it is an ARQ system, a special preamble has been added which allows for automatic tuning and very fast block synchronism.

To minimize the number of repeat requests, two alternative levels of FEC are added, which along with an option of no FEC allows for very flexible adaptation to conditions. Like MT63, the FEC uses multiple tones and interleaving, using time and frequency diversity to spread errors. The Q15X25 signal sounds a little like HF packet, with a preamble of tones, but follows with a burst of noise that contains the data. The signal is about 1800Hz wide. The Q15X25 protocol allows the use of AX25 or TCP/IP protocol stacks, so is ideal for packet radio. The performance far exceeds that possible with HF packet.

Timo Manninen OH2BNS has crafted the original SP9VRC code into a practical system for LINUX PC [4]. Unfortunately there is no Windows software for this excellent mode as yet.

Digital Voice Communications

Charles Brain G4GUO and Andrew Talbot G4JNT have shown that it is possible to transmit high-quality voice using a digital technique, in a bandwidth similar to analogue voice [5]. Charles uses a high-performance modem of his own design which transmits a signal rather similar to MT63 or STANAG 4285 – a wide band of what appears to be noise. Voice encoding and compression are used to achieve the necessary bandwidth reduction.

The modem signal is very robust and results from tests on 40m have been very encouraging. The modem uses parallel tone differential QPSK, 36 tones at 50 baud, and is capable of 3600bps. In fact 2400bps of data and 1200bps of FEC information are used. 50 baud is convenient as it matches the 20ms frame rate of the vocoder.

The first successful transmissions were in March 1999. Because of the FEC nature of the transmissions, they would be ideal for net and broadcast use. While the performance has improved as changes have been made to the system, there is unfortunately still a significant amount of specialised hardware required for digital voice, including a hardware vocoder (voice coder and decoder) and a special DSP modem system. The system is well described on Charles' website [6].

Ionospheric Sounding

While not a communications mode in the conventional sense, exploration of the ionosphere using sounding (like an HF radar) is an interesting pursuit for Amateurs. The pioneer of this technique for Amateurs is Peter Martinez G3PLX, recognised for the development of AmTOR and PSK31.

There are a number of ways to bounce signals off the ionosphere for the assessment of its height (delay) and other characteristics. Older sounders were just like conventional 'CW' radar, sending very high-powered pulses up into the sky, and measuring the delay of the resulting return. However, as with modern radar, sounding has progressed, and now many of the sounders are lower-powered frequency-swept or digitally-coded units.

Using DSP or other digital techniques, it is possible to compress the received echoes to enhance the specially-coded signal, and at the same time reduce the noise and sharpen the resolution. The details are well beyond this book [7]. There are many types of sounder. The type that caught Peter's attention was the 'chirped ionosonde', which transmits a very accurate swept signal progressing up in frequency at a very precise rate, and at a very precise time.

Commercial and military chirped ionosonde receivers are swept in frequency along with the transmitter, and so receive reflected components at different base-band frequencies, depending on the height from which the signal is returned. The receiver is an SSB receiver, and since the receiver frequency will have moved on by the time the returned signal arrives, the signal will not be zero beat, but will have an audio pitch dependent on the range. The receiver is there-

CHAPTER 16: ADVANCED DIGITAL MODES

fore connected to an audio-spectrum analyser that displays the responses.

G3PLX proposed a completely different passive approach, monitoring a single frequency (or narrow band of frequencies, the IF of a fixed frequency SSB receiver). The receiver searches for chirps of the correct type, those that swept upwards at 100kHz/sec. These sounders take just under five minutes to sweep the entire HF spectrum from 3 to 30MHz, and so can be found on just about any frequency.

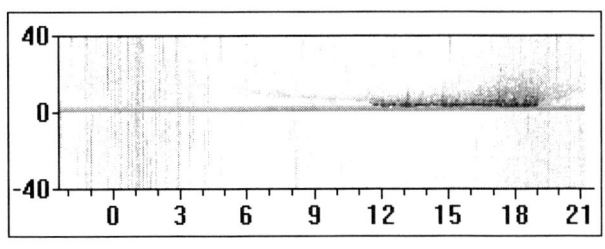

Fig 16.4: Ionogram with ground and sky wave. (Picture: Peter Martinez)

Using DSP techniques, a unique narrow filter was developed which only passed signals with the required upward chirping frequency characteristic. Software was developed which allowed the chirps to be logged from around the world, with a precise record of the arrival time, and to allow plots to be made of the resulting ionospheric reflections. Fig 4.15 and Fig 14.16 are recordings of this type.

The image in **Fig 16.4** was recorded on 10MHz by Peter G3PLX using the signal of a nearby sounder. Because of its proximity, the ground wave signal is detectable day and night (hence the straight line). The figure shows a strong short-delay sky wave during the afternoon, 1200 to 1900. There is also backscatter from the edge of the skip zone morning and evening. Notice how the apparent height of the reflective layer decreases as the signal becomes stronger, then increases as the skywave propagation dies out, leaving only the backscatter. The vertical scale of this image is in milliseconds, and the horizontal scale local time in hours [8].

The timebase accuracy required to produce these images is very high, for the signal to have no discernible slant over a 24-hour period. A drift or slant of 1ms over 24 hours would imply an error of one part in 86 million (about 0.01 parts per million).

The resolution of the G3PLX passive chirp sounding system is 125μs, quite sufficient to see fine details of propagation paths, and also sufficient to approximately locate the source of unknown chirp signals, given enough stations to measure the precise propagation time for triangulation purposes.

G3PLX has on numerous occasions been able to detect chirps that have travelled right around the world, even twice around the world. It is also possible for Amateurs to transmit and receive narrow-band chirp signals. It would be possible to advance or delay the chirps in order to add digital data (this is done commercially) so in theory it would be possible for Amateurs to send simple messages around the world using chirps.

Most of the passive sounding work has been achieved using specialised DSP equipment to process the signals for PC logging and display [9]. Two special items were required – a Motorola DSP56002EVM development kit, and a very precise 1Hz time reference. Most users have chosen to use a modified GPS receiver to provide the time reference. This is because very high-precision time references are expensive, and still leave the user with the question of synchronising time with other observers. GPS solves these problems at modest cost.

Because of the difficulty in obtaining the necessary equipment, until recently the number of users was limited to less than a dozen. Now, software for processing the signals using a PC and sound card has been developed by Andrew Senior G0TJZ. This software, called Chirpview, uses a GPS unit with 1Hz output as the precision time reference [10]. This puts chirp sounding within the grasp of any Amateur.

References

[1] See http://www.chbrain.dircon.co.uk/pcale.html
[2] See http://www.chbrain.dircon.co.uk/hfemail.html
[3] See http://www.tapr.org/tapr/html/Fdcc99audio.html
[4] See http://www.peak.org/~forrerj/HISPEED/index.html
[5] See 'Practical HF digital voice', Charles Brain G4GUO, *QEX* May/June 2000, also available at http://www.arrl.org/qex/brain.pdf
[6] See http://www.chbrain.dircon.co.uk/dvhf.html
[7] The techniques are well described in the article by Martinez, 'The chirps project: a new way to study HF propagation', *Radcom*, July–August 2000.
[8] This and several other fascinating images are described in the previously referenced article (see Footnote 7).
[9] This software is available from http://www.qsl.net/zl1bpu/chirp/chirps.html
[10] See http://www.asenior48.freeserve.co.uk/chirpview.html.

17

Other tools for amateurs

In this chapter
* Spectrograms and other spectrum analysers
* Dopplergrams
* Oscilloscopes
* Wave tools
* Other analysis tools

THE PC with sound card has become an impressive tool for Amateur use. The author has been widely quoted as suggesting that the PC spectrogram is the most important invention for the ham shack since the grid-dip meter. There are a number of other more- or less-specialised tools – oscilloscopes, wave analysers and meters of various types that are equally useful.

Spectrograms and Spectrum Analysers

Both of these tools are generated using a Fast Fourier Transform, a mathematical process. A spectrum display simulates a spectrum analyser, showing signal strength (vertically) and frequency (horizontally). The sound card is a perfect tool for this application, and is able to operate from close to DC up to 20kHz, and indicate signals with a dynamic range of over 60dB. There are many simple spectrum displays in this book.

The spectrogram is a graph of signal frequency versus time, and is generated in the same way. In this case, however, the signal frequency is one axis (typically horizontal) and time is the other axis (usually vertical). Signal strength is portrayed as brightness. The spectrogram is the perfect tool for analysis of digital mode signals. Displays of this type are often called waterfall displays because of their similarity to a waterfall in appearance. For long-term frequency monitoring (drift etc) the display is usually portrayed sideways, with the time axis horizontal. All the spectrogram images in this book are of this type.

'Spectogram' by Richard Horne [1] is perhaps the most frequently-used tool for spectrums and spectrograms. This program is very flexible, offering a wide range of speeds and resolutions, averaging, colour and black-and-white displays. Most of the spectrums and spectrograms in this book were generated with this program. The special radio version allows the FFT conversion display rate to be controlled

Fig 17.1: SpecLab in 3D mode displaying CW on 20m

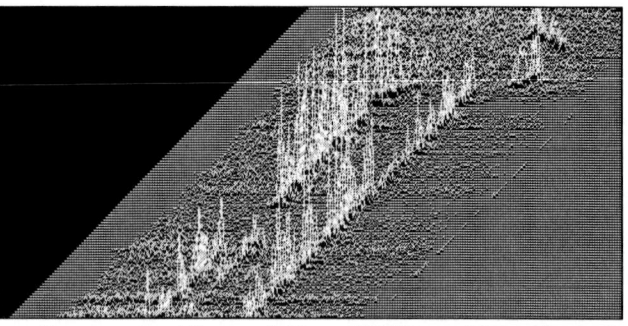

precisely. Unlike other programs, the intervening conversions are not thrown away, but are averaged, reducing noise and improving sensitivity. Other suitable programs are Spectran by Alberto di Bene I2PHD and Vittorio de Tomasi IK2CZL [2], and SpecLab by Gerhard Wiche DL5NDH (**Fig 17.1**) [3].

SpecLab has interesting three-dimensional spectrogram displays (like a combination spectrum and waterfall display) and many other tools. Unfortunately the documentation and controls for SpecLab are at present only available in German.

Dopplergrams

Dopplergrams are a specialised type of spectrogram that offers a very slow sampling rate and a very narrow bandwidth. The most common use is for monitoring Doppler effects, such as on HF carriers or meteor scatter, but they are also useful for the adjustment of frequency standards and monitoring drift of receivers or transmitters. For high-resolution Dopplergrams, a very stable receiver is required.

The best specialised Dopplergram software at present is EVMDop by Peter Martinez G3PLX. While it operates with a Windows display, unfortunately this program requires special hardware, the DSP56002EVM digital-signal processing kit. This unit is quite versatile, and there are many different programs for the well-known 'EVM', even the very first software for PSK31 and MT63, but unfortunately the unit is no longer available. Fig 4.17, Fig 4.18 and Fig 10.8 were generated using EVMDop. It gives the sharpest pictures of all the programs available.

Spectran by Alberto di Bene I2PHD and Vittorio de Tomasi IK2CZL can be adjusted to provide narrow band spectrograms suitable for some Doppler use. Spectran is intended for EME applications (moon-bounce) and has a detector optimised for weak signals (see **Fig 17.2**), so it is not ideal for this application (the high sensitivity means that the images are not as sharp). The resolution is good however, with spans below 10Hz easily achieved.

Another program by I2PHD and IK2CZL is suitable for some Dopplergram work, and is much easier to use. ARGO is designed for LF weak signal reception, especially receiving very slow Morse [4]. In 30-second mode it has a span of only 6Hz. Unfortunately the resolution of the image for Dopplergrams is limited by the special weak signal detector, perfect for weak signals, but not for Dopplergrams. ARGO can be used to make quite acceptable meteorgrams and is useful for drift monitoring.

Oscilloscopes

There are two simple oscilloscope programs of interest to Amateurs. The best is WinScope by Konstantin Zeldovich [5], which has dual-channel operation, an

CHAPTER 17: OTHER TOOLS FOR AMATEURS

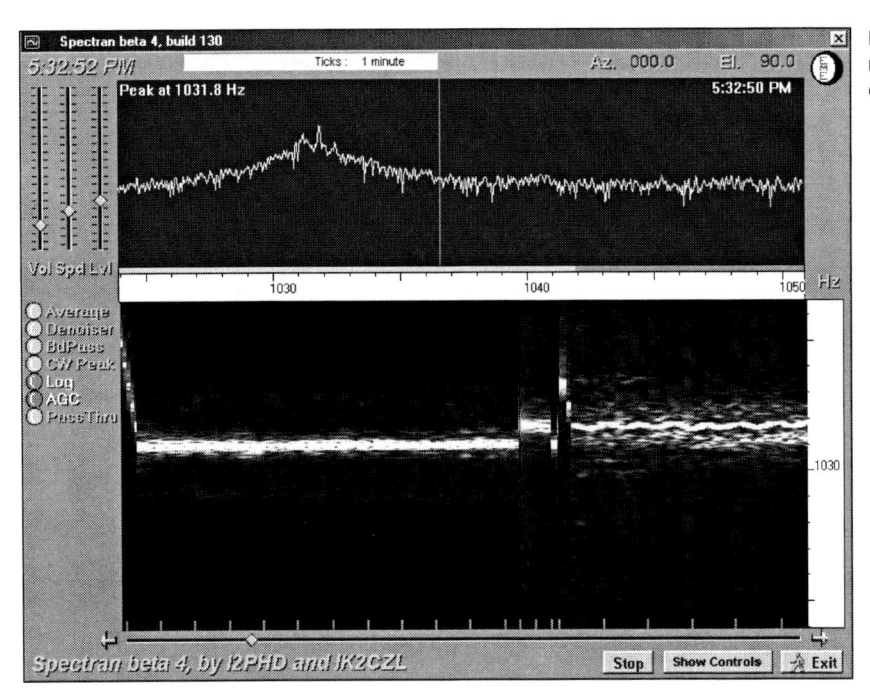

Fig 17.2: Spectran monitoring a weak carrier

X–Y mode and excellent triggering. It operates very like a conventional oscilloscope. The other program, by BIP Labs, is simpler. BIP Labs also offer other simple sound-card tools [6].

While the oscilloscope is useful for real-world signals, it is not as useful for analysing synthesised waveforms as a wave tool, which is more like a storage oscilloscope.

Wave Tools

These complex and powerful programs are intended for manipulating digitally-sampled sound files, typically .wav files. Most can also record and play files, and in this context the ability to view small samples of files is very useful. For example, the samples of PSK31 and PSK-Hell waveforms in this book, eg Fig 13.6, and Fig 9.17, were created using a wave tool, GoldWave [7].

A good wave tool also allows sound samples to be manipulated and mixed, allows sounds to be generated mathematically, or changed in frequency, sample rate and signal level. With these tools it is possible to simulate different ionospheric conditions to assess the performance of digital-mode receiving software. Fig 9.16 was created in this way, by adding noise and simulated Doppler shift to a PSK-Hell signal recorded in GoldWave. Sound Forge [8] and Cool Edit [9] are among the many other wave-tool programs with suitable attributes for Amateur applications.

Wave tools and some of the spectrum analysis software also allow very precise measurement of audio frequencies. This can be of great benefit when measuring the frequency of a transmission accurately. The receiver-offset error can be checked by carefully measuring a reference-frequency source, and then the offset error is added to the sum of the receiver display and the measured audio frequency.

Fig 17.3: The SpecLab auto-correlation analyser

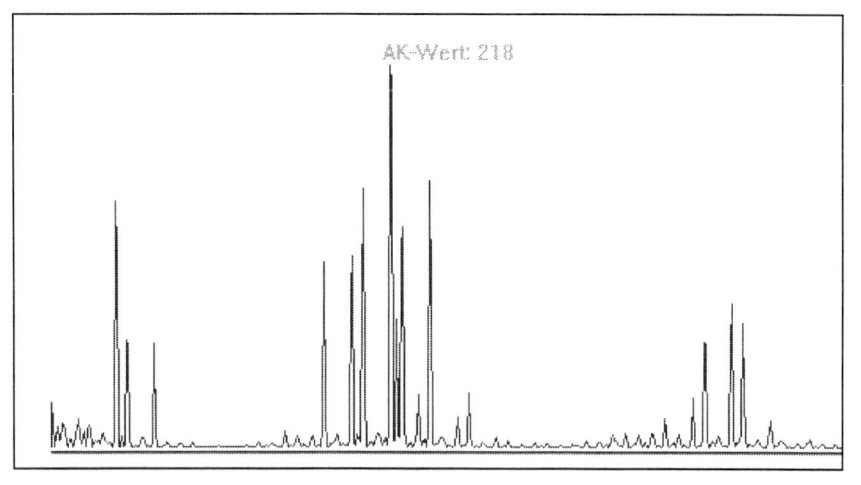

Other Analysis Tools

There are many different applications that can be envisaged with a PC sound card. For the Amateur, especially one new to digital modes, some means of identifying the sounds heard is perhaps the most pressing need. It is for this reason that there are many standard resolution spectra and spectrograms in this book. These provide the best clue to the nature of a signal, and are easy to replicate.

For further analysis, for example to determine the baud rate of a signal, or to determine whether a PSK signal is 2-PSK or 4-PSK, other tools are needed, for example an autocorrelation analyser for baud rate analysis, and a phasemeter or constellation display for phase analysis.

These tools are available for the PC sound card, and can be simple to use, but are much more difficult to interpret correctly. There are many other useful tools as well:

♦ Phasemeter – uses a constellation diagram to identify PSK modes and determine how many phases exist

♦ Correlometer – auto-correlation is used to compare the signal with a delayed version of itself. Strong auto-correlation is found at the baud rate

♦ Data scope – displays demodulated data

♦ Crossed ellipse – useful for tuning FSK modes

♦ Eye display – useful for assessing demodulator performance and signal timing distortion

♦ Distortion meter – displays timing distortion in teletype signals

♦ Power meter – displays average and peak power of signal, and can assess fading

♦ Signal scope – displays a modulated signal for assessing AM, noise and fading

♦ Data-pattern display – can help discover baud rate, synchronous modes and patterns

Many of these analysis tools are available in SpecLab (**Fig 17.3**) and Skysweeper (**Fig 17.4**) [10].

References

[1] See www.monumental.com/rshorne/gram.html
[2] See http://www.weaksignals.com/
[3] See http://people.freenet.de/dl5ndh/SpecLab.html
[4] Also at http://www.weaksignals.com/
[5] See http://www.mitedu.freeserve.co.uk/Prac/winscope.htm
[6] See http://www.stud.fh-hannover.de/~heineman/freeware.htm
[7] See http://www.goldwave.com
[8] See http://www.sonicfoundry.com/products/default.asp
[9] See http://www.syntrillium.com/cooledit/
[10] See http://www.skysweep.com

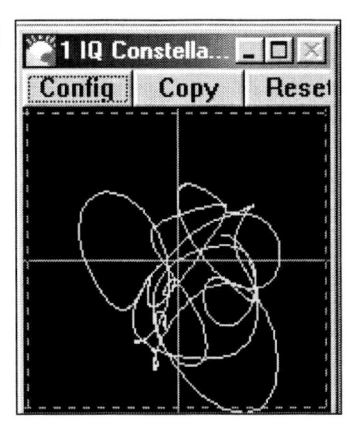

Fig 17.4: The Skysweeper PSK31 constellation display

A

Digital modulation techniques

In this appendix
- The ASK modulator
- Complex modulation
- Data synchronism

MANY AMATEURS are apparently interested in learning how digital modes work, in terms that are simple to understand. This is by no means easy to do. In Chapter 2, an overview of the different techniques was given, with very little detail. This Appendix is the place for more detail, hopefully still understandable by those with little knowledge of radio or digital-electronics theory.

As a means of assisting understanding, the three main modulation types will be described, and for each one, a typical example will be given where we can follow the course of each key press through the software, out to the computer, and over the air.

Later in the chapter is a discussion of synchronising techniques that is more detailed than the beginner requires, and yet is useful for reference purposes.

The ASK Modulator

This is probably the simplest modulation technique of all to understand. It is part of the Amplitude Modulation (AM) family, simplified for Digital Modes. Basically, the carrier is keyed on and off, or increased and reduced in power.

For Amplitude Shift Keying (ASK), also known as On–Off Keying (OOK), a number of modulator designs are suitable, for example direct switching of an amplifier-stage power supply (typical of CW Morse) or using a simple balanced modulator.

The balanced modulator uses a signal on one input port to adjust the amount of output of the carrier at the other port. When switching on and off an amplifier, there are normally only two possible amplitudes – full output and no output. With a balanced modulator, and with some other techniques, it is possible to arrange for any two different carrier levels by altering the bias of the modulator.

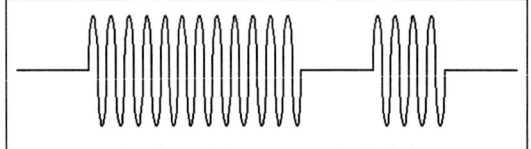

Fig A.1: An ASK example – Morse

ASK is used for Morse (**Fig A.1**), Hellschreiber (Feld-Hell), and transmission of time-standard signals such as MSF or DCF77. Despite the simplicity, ASK is not widely used for digital modes on HF because it is difficult to receive accurately when there is noise or signals are varying in strength.

In Fig A.1, imagine the letter 'N' is sent in Morse. The sine-wave carrier is turned on, then off, for the duration of a dash, then again for the dot. When the carrier is off, there is no signal, so this is 'on–off keying'.

When generating ASK in software, digital signal processing techniques (DSP) are used, and the subcarrier is generally created as a sine wave from a table of numbers. Less computer power is required with this technique than calculating the sine-wave values as they are required. To alter the amplitude of the signal, the value from the table can be multiplied by 1 (for on) and by 0 (for off), or by some value in between.

ASK Example – Hellschreiber

When you press a key on the keyboard, the ASCII value for the key (simply a number from 0 to 127) is stored in a buffer, a specially-operated type of memory. The buffer memory is operated in a 'first in, first out' manner (FIFO), rather like water entering a hose pipe – the first water in is the first out at the other end. Characters can be added to the buffer during transmission, or even while the transmitter is not operating.

When the transmitter is operating, it gets the next character from the buffer (the oldest), and looks up the code to be transmitted for that character. In the case of Hellschreiber, this consists of several bytes of data describing the dots to be sent. For example, to send the letter 'A', the bytes might contain the following (expressed in binary for easy visibility):

```
0000000
0010000
0101000
1000100
1000100
1111100
1000100
1000100
```

In this example there are seven bytes, each of course containing eight bits. The bytes are drawn vertically. Look closely and you will see the letter 'A' traced out by the ones. This is what has to be transmitted. The bytes are transmitted from the bottom left to the top left, ie the bits of the first byte in order, then the second, and so on (remember the bytes are drawn vertically in this example). The last two bytes contain no transmitted data, and depending on how clever the software is, they may or may not be transmitted. Most software uses this information for 'proportional spacing', ie wide characters like 'W' have more columns (more bytes) than narrow letters like 'I'. There is always a delay equivalent to two columns between characters anyway.

So, for each character in the buffer, these codes to be transmitted are recovered from a fixed area of memory (called a table) that defines the font to be

APPENDIX A: DIGITAL MODULATION TECHNIQUES

used. The transmitter then has to control the transmitter using an AM technique. In the case of Feld-Hell, each '1' is converted into a raised cosine shape (like an inverted 'U'), which then AM modulates the carrier (see **Fig A.2**). This gives a clean signal with minimum bandwidth. If a square-keying signal were used, they would be strong key clicks.

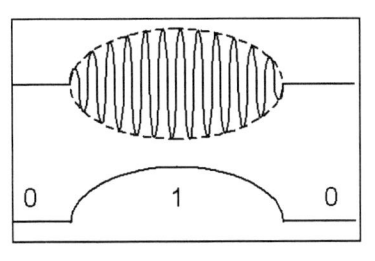

Fig A.2: Raised cosine modulation of Hell dots

It is easy to imagine the generation of this smooth dot shape by a conventional AM modulator. Well, it is just as easy with a balanced modulator or a software modulator. Both consist of multipliers (mixers), and the shaped dot is simply multiplied by the sine-wave carrier to produce the shaped envelope. When the shape is zero, there is no output: when it is at a maximum, there is full output.

So now we have each one in the data converted to a shaped dot (in software). The data is simply shifted out to the sound card at the chosen sampling rate (typically 8000Hz). To make this easy, the carrier and the raised cosine shape are both defined in terms of 8000Hz samples. For example, since Feld-Hell transmits at 122.5 baud, each dot will contain 8000/122.5 = about 65 samples.

The sound card operates as a digital-to-analogue converter. Each of the samples sent to the sound card is a 16-bit number (the result of all the mathematics), and they are sent to the outside world at exactly the correct sample rate, so the data rate and signal frequencies will be accurate and constant. Fortunately the sound card is connected to the computer processor by a large buffer, so even if the processor operates in jerks, which it may if it is busy doing other things part of the time, the data will come out of the sound card smoothly.

After the digital-to-analogue converter is a low-pass filter (which removes the sampling frequency noise) and a simple audio amplifier. The sounds (which are just as you would hear them in your receiver) are then sent to the transmitter audio system for transmission. Since the transmitter is a linear SSB transmitter, the audio signal is simply translated up in frequency to the spot on the band to which the receiver is tuned. Since the low-pass filter in the sound card is designed for the full range of the sound card (up to 20 kHz), it is important for the sound sent to the transmitter to be limited to the communications voice range at least. Fortunately the transmitter audio shaping circuits and crystal filter will do this adequately.

ASK Example – Morse

Transmitting Morse from a computer is very simple. The ASCII value for each character pressed on the keyboard is stored in a buffer, and retrieved in order during transmission, just as with Hell and RTTY. The character is then looked up in a table (see the Morse table in Appendix B). The value from the table is just one byte (eight zeros or ones), and contains all the necessary dots and dashes in order. Dots are stored as zeros, and dashes as ones. The spaces in between are not stored at all, since they are added using a rule.

Let's watch what happens when the character 'C' is sent. The value in the table for 'C' is 15_{16}, ie hexadecimal 15. This can be represented in binary as:

00010101

Read the first four binary digits from right to left, and the result is:

▬ ▪ ▬ ▪

which is, of course, 'C'. The rules used say that the binary bits are shifted right until the result is 0000001 (01 hexadecimal), sending each bit that shifts out of the byte, and replacing each bit at the left with zero. This might sound complicated, but is extremely easy to do with a computer, and also allows even the most complex Morse character to be stored in one byte. Watch what happens with the letter 'C':

00010101	Ready to send
00001010 1	First shift. First bit sent is '1' or 'dash'
00000101 0	Second shift. Bit sent is '0' or 'dot'
00000010 1	Third shift. Bit send is '1' or 'dash'
00000001 0	Fourth shift. Bit sent is '0' or 'dot'

Since the byte now contains 00000001, no more need be sent, and the spaces following the character can be sent. Dots and dashes are always followed by silence of one dot time. The end of each character is always followed by an additional two-dot-times silence (sometimes three are used), and whenever the space character occurs, silence of five dot times is sent.

Dots and dashes are sent using a sound card by sampling a sine-wave table at a fixed rate (the sound-card sample rate) to make a fixed-pitch tone. The number of samples sent depends on the Morse speed. Silence is achieved by sending the same number of samples, but all of the same value (resulting in no sine wave). Very often a raised cosine or other filtering technique is used to round off the edges to prevent key clicks. It is also helpful if the sine waves start and finish at a neutral value that is the same as the value for silence, to eliminate any 'thump' that might occur as an element starts or stops.

To send Morse with a small microprocessor, an audio oscillator (which may be in the microprocessor) is directly keyed by a microprocessor output, and drives an SSB transmitter. Sometimes a CW transmitter is directly keyed from a microprocessor pin. The microprocessor maintains a timer to time each dot, and the output is simply raised for dot or dash, and lowered for silence. Most beacons use this technique.

The FSK Modulator

Certainly the most widely used modulator for digital modes, the FSK (Frequency Shift Keying) modulator is used to change the frequency of the transmitted carrier. Of course FSK is a digital version of Frequency Modulation (FM).

Because the modulator works on the carrier frequency, when generating FSK with conventional hardware techniques, the modulator acts directly on the carrier generator. When a subcarrier is used, the subcarrier oscillator can be frequency modulated or a choice made between two oscillators.

You can see in **Fig A.3** that when the datum (lower square-wave trace) is zero, the carrier frequency is low, and when the datum is one, the frequency is high. For the signal to be clean, it is important to avoid sudden changes in phase when the audio tones are switched. If a single oscillator is used, it may naturally maintain constant phase when changing frequency. In Fig A.3, the frequency changes as the sine wave crosses the zero point. It does not matter where the frequency changes provided the phase is the same at each frequency at the transition points.

When generating FSK in software, a numerically-controlled oscillator is used. This also has the interesting property of frequency shift without phase shift, so the signal generated can be very clean. To change frequencies, the generator simply uses a different number of steps for each frequency.

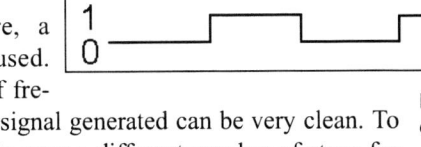

Fig A.3: An FSK example

The spectrum of an FSK signal is like two (or more) ASK transmissions sent alternately, so there is always transmitted power on one frequency or another. The advantage comes in always have a signal plus noise energy at one frequency to compare with the noise level at another.

MFSK is just the same as FSK (2-FSK) except there are more levels to the data (not just one bit, zero or one), and therefore more frequencies are generated.

FSK Example – RTTY

As with Hellschreiber, characters typed on the keyboard are stored in a buffer as ASCII numbers until needed. In the case of RTTY, only one byte is needed per character, so when it is time to transmit each letter, the code lookup uses a table containing the ASCII to ITA2 conversion, one byte per character. It is normal to store the required shift with the ITA2 value in the lookup table.

Because the ITA2 transmission has a shift system, an extra function is necessary. The software must keep track of the current shift (LTRS or FIGS), and if the ASCII character read from the buffer is the other shift, a shift character (FIGS or LTRS respectively) must be transmitted first, so whenever a shift change is needed, two characters are sent to the transmitter.

The output of the lookup table has only five ITA2 bits per character. When the character is transmitted, a fixed one-bit duration start pulse is added (always one), then the five data bits are sent in order, then a stop bit is sent (always zero). The stop bit must be at least 1.5 bit times long, and the next character cannot be sent until this is completed. If no characters are ready to be sent, the stop bit carries on forever (idle). See Fig A.8 later in the chapter.

These ITA2 data bits to be transmitted are used to change the step size in a numeric oscillator, just as they would be used to change the frequency of a real audio oscillator in a conventional RTTY TU. The result is an audio tone at one of two different pitches.

At 45.45 baud, each data bit is 22ms long, so to ensure that the sounds are created at the correct length by the sound card, an exact number of samples must be sent for each bit, since it is this process and the accurate sampling rate of the sound card that defines the speed.

For convenience, the numerical oscillator used to create the tones operates at the sample rate, or an exact proportion of it. For example, if the numerical oscillator produced tone samples at 8000Hz, there will be about 176 samples per bit.

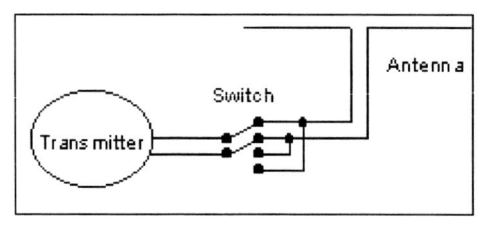

The PSK Modulator

PSK is perhaps the simplest modulation in concept, and yet it is the hardest of the three common modulation techniques to understand. PSK (Phase Shift Keying) is very simple to generate.

Imagine a CW transmitter operating into a dipole antenna fed with open-wire line. Imagine also that we have a double-pole switch in the feed line that effectively swaps over the feeders. It won't matter which position the switch is in, the carrier will be the same strength, and the signal will consist of a single carrier. **See Fig A.4**.

Imagine now that we were to switch the feed line over, back and forth, many times per second. At one moment the RF field will have one phase, at the next the opposite phase.

This change in phase is easily detected at the receiver, by comparing the phase at one moment with the next, using a local stable reference. In this way it does not matter what the actual transmitted phase is, nor the actual phase of the local reference.

PSK is now widely used because it is a sensitive technique that also uses little bandwidth. PSK is easily generated by simply changing the way a numerically controlled oscillator is stepped, by using a switched inverting amplifier, or by changing bias on a balanced modulator.

In **Fig A.5** the phase changes at the zero point, but because the effect is the same as turning off a transmitter of one phase and turning on a transmitter of another, with no shaping, there will be very strong keying transients (even stronger than the clicks generated by CW transmissions). These can be removed by filtering, or by shaping the amplitude of the carrier so it is at a minimum when the phase is switched.

Remember the raised cosine used for Feld-Hell? The same technique works here, except that the modulating waveform has positive and negative peaks, one for '1', the other for '0', one phase and the other. A balanced modulator or software equivalent will perform this trick if the data is pre-shaped as a raised cosine shape as illustrated in **Fig A.6**. (In Fig A.8, later, the phase changes when the signal power is at a minimum, controlled by the shape of the data waveform.)

To avoid uncertainties caused by drift and ionospheric effects, no particular phase is considered to be '0', and another '1'. A differential PSK technique is used where for example sending one phase then the opposite phase might mean '0', while sending one phase and the same phase again might mean '1'. So long as the drift in phase (of the receiver, local phase reference and phase changes in the ionosphere) is small during the reception of each data element, the data will be interpreted correctly.

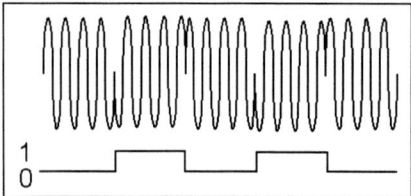

PSK Example – PSK31

It is a little daunting to try to explain PSK31 in layman's terms, but here goes. As with the other modes described, the keyboard characters are stored in a buffer, and retrieved

Fig A.4: A simple PSK transmitter

Fig A.5: A PSK example

as necessary. The transmitter timing is fairly critical, and the transmitter must always be transmitting something in order for the synchronising signal to be available at the receiver; so when no data is available a series of '00000' bits is sent. Now, since the PSK31 modulation is differential PSK, to send a zero means sending a different phase for the next bit. Fig A.6 is a bit confusing, in that it would seem that the data sent is '101', but in a differential PSK mode this is the sequence for '000'.

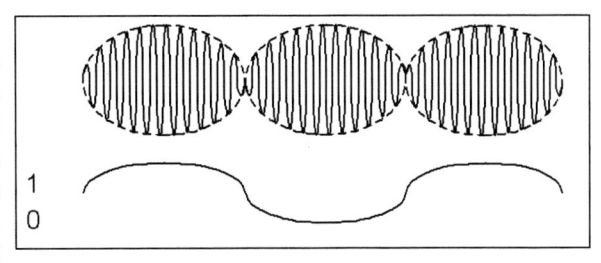

Fig A.6: PSK with raised cosine envelope

When data is available for transmission, the software will read the next character from the buffer and look up the character in a Varicode table (which will contain from one to ten bits for each character). The software will then set about transmitting the bits. If the bit to be sent is a zero, a one or zero is sent to the transmitter modulator in order to change the phase – if the last phase was one, the next one will be zero. If the bit to be sent is a one, the phase sent to the transmitter will be the same as for the last bit. The bits are shaped as a series of samples to generate the raised cosine shape, as previously described.

At the end of each character, a '00' sequence is transmitted. This signals to the receiver that the character has ended, as no Varicode combinations contain '00'. This is also convenient, as the idle signal consists of a long train of '00' sequences.

In PSK31 all the bits are the same size, exactly 32ms long. With an 8000Hz sampling rate, this is very conveniently exactly 256 samples, so in fact it would be possible to store a complete data bit as samples, rather than calculate the waveform. This is an important factor for slow computers. There would be only one table of 256 samples for a bit required, as the opposite phase could be transmitted by sending the samples in reverse order. The samples are sent to the sound card to be converted from digital numbers into an audio waveform for transmission. Since each bit always has the same number of samples (256 in the example), and the sampling rate of the sound card is very accurate, the tone frequencies and the bit-rate timing of the signal transmitted will be very accurate.

Real PSK31 software does not use a fixed table to define the shape of the tone burst of each bit, as this would only work at a fixed frequency. Instead, once again a numerically controlled oscillator is used and the output modulated by the raised cosine envelope and data stream. The pitch of the audio tone is defined by the step size used in the NCO.

Complex Modulation

Very often several modulation techniques are used at the same time. As mentioned above, it is not uncommon for amplitude modulation and phase modulation to be used together, in order to ensure that at the point where the phase is changed the amplitude is at a minimum. The modulation usually has the appearance of a sine wave 100% modulating the carrier, and is achieved by using an additional balanced modulator where one modulating signal is carrier, and the other a raised cosine wave, which consists of a sine wave (or cosine wave) superimposed on a DC component (**Fig A.7**).

Other shapes can be used for the same purpose, with differing keying bandwidth properties [1].

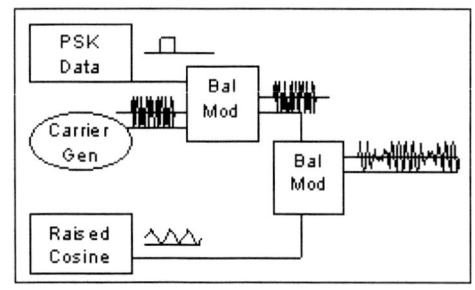

Fig A.7: Raised cosine modulation

Other modulator combinations are possible. Clover is one of the most complex. It has four carriers, each separately amplitude modulated and phase modulated, and each transmitted at a different time, so it combines all three modulation techniques. This type of signal is frequently called an Orthogonal Frequency Division Multiplexed (OFDM) type, because the three modulation techniques are arranged so that the amplitude, frequency and phase data can always be recovered independently. Some other Amateur modes (MT63, Q15X25) and many of the more powerful commercial and military modems use an OFDM technique. An intriguing thought is that very often in complex modes the signal is generated not by software modulators as described above, but by describing the signal as frequency, phase and amplitude components, and then performing a 'reverse FFT' mathematical calculation, like a waterfall display in reverse. Why not, since they are often received using an FFT?

Data Synchronism

This section is for those who wish to know more about how transmitted data is kept in synchronism at the receiver, ie how the receiver can be sure that each data bit is received and sampled at the correct time. It is not a subject often considered by Amateurs, but the type of synchronism used has a major impact on the performance. Some techniques are very efficient; some are very robust; and all modes involve some sort of design compromise because of the synchronising technique chosen.

Character Asynchronous Transmission

The best-known mode of this type is RTTY. The start of each character is indicated by a start event which is relatively unique, followed by a series of accurately-timed data bits, where the data bits are timed from the start event, and a relatively unique stop bit. Since RTTY is so universal, this mode will be used to describe the technique in detail.

The signal is sent on a single wire or on a single radio channel in a serial manner, so there are two possible states, historically known as 'mark' and 'space'. These names are well over a century old, and relate to when early Morse equipment made a mark on a paper strip (when current was flowing in a solenoid) or left a space (when no current flowed). Typically in RTTY equipment, the mark condition (current flowing) is the idle condition. When the channel is idle, the mechanical equipment or its electronic equivalent is idle, waiting for a start event.

Fig A.8 illustrates the condition of the data line with passing time from left to right. The voltage levels are shown as they would be on an RS232 line, −10V for mark, and +10V for space.

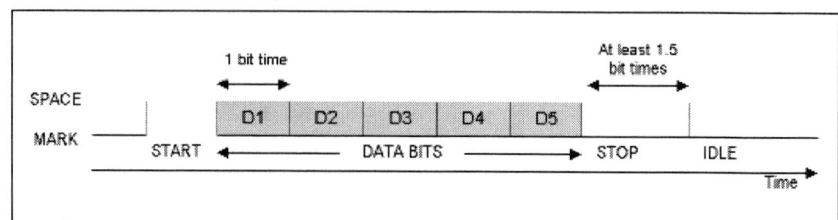

Fig A.8: Asynchronous Character Transmission

APPENDIX A: DIGITAL MODULATION TECHNIQUES

The idle condition prior to start terminates with the arrival of the start bit, which is the same length as the data bits, but is always a '1'. At the start of this bit, which is typically 22ms for 45.45 baud or 20ms long for 50-baud transmissions, the receiving equipment starts timing to receive the data bits. In a mechanical machine a clutch is released and a shaft starts to rotate at a constant speed. In an electronic system, a counter is started, ready to clock data into a shift register. This point, the start of the start bit, individually synchronises each character, and the subsequent data bits are sampled exactly in the centre of each data bit.

In a mechanical machine, this sampling process moves five small pins or levers via a clever 'pecker' mechanism steered by the magnet; in an electronic system the data is sampled into a shift register by the data clock. If there is a slight difference in the sending and receiving speeds, this sampling point will slowly drift away from the centre of the data bit as the later bits of the character are received. Because the character reception is synchronised at the start of each character, the speed difference needs to be significant (rather greater than 1%) before errors are likely.

Following the five data bits (RTTY uses the 5-bit ITA2 character alphabet), a stop bit is sent. This is always a mark bit ('0'), but in order to help make the synchronising process unique, this bit is always longer than the other data bits (typically 1.5 to 2 times). During this time the electronics returns to idle (or the machine shaft stops) and the character received is transferred to the next stage of the receiver.

The stop bit is followed by an idle condition, which is indistinguishable from the stop bit. The idle period is of indeterminate length, and can be as short as zero. When text is sent at full speed this idle time will be zero, and represents the fastest speed at which the circuit can send data.

The stop/idle is always mark; the start bit is always space, and the start always follows a stop/idle, so the equipment can readily distinguish where to start timing. This is important so that reception can start correctly when the equipment starts in the middle of a transmission, or if there is a burst of noise, interference or fading which disrupts the reception. The technique is not perfect, as this sequence can from time to time be mimicked by the data (eg data bits '001'), and so frequently there are several characters printed in error before correct synchronism is attained.

Mechanical RTTY machinery performed remarkably well when correctly adjusted. Modern electronic systems use more sophisticated sampling of the data, and so can better recognise distorted or noisy characters accurately. In addition, the timing in electronic systems is typically very precise and the start and stop bits are recognised more reliably, so performance is improved and adjustments are eliminated.

Communications between computers using an RS232 or RS422 communications connection work in exactly the same way. Computer data is usually ASCII or 8-bit data, and is sent with one start bit, seven or eight data bits and one stop bit. Speeds as high as 38,400bps are common over short distances. Because the data is sent on wires as DC levels, not tones, it is incorrect to talk of 'baud rate' as there are no symbols to time, only bits (baud rate means symbol rate). Next time your friend tells you his serial printer or TNC communicates with his PC at 9600 baud, you can correct him – he means 9600 bits per second!

Character Synchronous Transmission

The most important difference between this technique and the asynchronous technique just described is that there is no steady idle condition. Data bits are sent continuously, and when data runs out (typing stops momentarily or the end of a file is reached), something has to be transmitted to keep the system happy and in bit synchronism. This is of course called an 'idle bit' or 'idle sequence'.

In this type of mode, character synchronism is achieved by sending a unique bit sequence (ie one which cannot appear in the data) as the means of indicating when a character starts.

This means that generally a larger number of bits must be used per character, since many of the possible combinations will not be permitted. For example, imagine that '0001' was chosen as the character sync signal. Then data words '00001', '00010', '00011' would not be permitted since they contain the synchronising signal.

This synchronous technique can be very robust, as the synchronising signal can be repeated over and over at precise intervals and can be extracted with great precision even when the signal is very noisy. (This is an auto-correlation technique.) Thus a system of this type can have significantly improved noise performance, as the synchronism is hard to disrupt. AmTOR mode B can benefit from this type of sync extraction.

Character-synchronous transmission must use an independent technique to discover or maintain the bit synchronism, ie when to sample each bit. One common method (and there are several) is to transmit each bit as three events, like a tiny asynchronous character with one data bit. For example, '100' might mean '0' and '110' might mean '1'.

Examine for yourself how a sampling system synchronising at the '1' at the beginning of the bit's first event, and sampling in the centre of the second of the three events, will be able to correctly sample the data. This is called a 'self clocking' technique. Two other techniques are common; one where data is transmitted differentially, so a change from zero to one or vice versa at the start of a bit not only indicates when the bit clock occurs, but the state of the next bit. Of course the state will often need to change during the bit in order to signal the correct phase at the next clock transition. The other common technique is to amplitude modulate the bits (as in PSK and some MFSK modes) so that the bit clock can be recovered.

The simplest examples of this type are AmTOR modes A and B. AmTOR uses a special code, the Moore code, which has five data bits (as in ITA2) and two parity bits, arranged so that all valid characters always have four mark bits and three space bits – any other combination is not permitted. The seven-bit code allows for all 32 ITA2 codes and three special control characters. These control characters are transmitted in a unique sequence to allow synchronism to be achieved.

SSTV is a synchronous image mode. In this case, the 'character' is a line of analogue video data, and the sync signal is a pulse at the start of each line. SSTV can also benefit from correlated sync extraction.

Bit-stream Synchronous Transmission

What happens when the data transmitted is binary, rather than characters, or consists of alphabet characters with different lengths? The system just described still works. The receiving system simply has to handle characters of different sizes, and expect the synchronising signal to arrive at unpredictable times to terminate the character.

PSK31 is a transmission of this type. Bit timing is recovered from the envelope of the transmitted signal, but characters are identified by a unique bit sequence '00' that is used to separate them. In order to handle a condition of idle transmission, which consists of repeated '0000' bits, the sequence used to separate characters is defined as any sequence of more than one '0'. The information sent in PSK31 and similar modes can be binary data or characters with from one to ten or more bits per character.

Block Synchronous Transmission

Perhaps the most commonly known mode of this type is AX25 packet. Like other synchronous systems, a unique bit pattern is used as the synchronising signal, but in this case it synchronises not just a character, but a whole block of many characters, quite appropriate when conditions are stable and timing is very accurate. In this system, the unique sequence cannot be allowed to exist inside the message, so when this condition is discovered in the data to be transmitted, an extra bit must be added, and of course removed at the other end, a technique called 'bit stuffing'.

Imagine that the synchronising signal for the block was '1001'. This is sent to mark the start of the block, which could be any length, and then the data is transmitted. When the sequence '1001' is found in the data, imagine that the final '1' is replaced by the data '01'. This is called 'bit stuffing'. The receiving software needs to be able to recognise both '1001' (the synchronising sequence) and '10001', which it replaces with data '1001'. The receiver also needs to be able to recognise a genuine un-stuffed '10001' sequence, so such systems need to be designed with considerable thought.

Systems of this type typically commence transmission with a sequence such as '0101010', which is used to synchronise the data-bit timing before the block sync pattern just discussed. Bit timing needs to be very accurate as it must remain synchronous for the full length of the block transmission. Systems such as this are generally used on VHF where noise is minimal and timing errors are small. Block synchronous transmission is very efficient, since the synchronising data is only sent at the start of a block, rather than for every character.

HF-FAX is a block-synchronous image mode. The sync signal is transmitted as a sequence of distinctive pulses, only at the start of the picture, and the rest of the picture is then received using the accurate and now synchronised clock.

Quasi-synchronous Systems

When a system is designed so that the received data need not be sampled in synchronism with the transmission, but is sampled independently at about the transmitted rate, synchronism need not be very precise. The term 'quasi-synchronous' specifically applies to modes that have a notion of synchronism, but where the design of the mode allows the receiving system a wide margin of speed and sampling phase error before reception becomes difficult. The 'fuzzy' modes are generally of this type.

Hellschreiber is the best known example of a quasi-synchronous mode. The receiver is arranged to print each character twice, so it does not matter where in the character transmission the printed text starts, as it will always be printed clearly at least once. The bit sampling in most Hellschreiber modes is non-synchronous, as it does not matter when the text is sampled, providing it is sampled often enough, as the eye 'joins the dots' to reconstruct the characters. The Hell transmission and reception systems can have completely different transmitted bit rate and receiver sampling rate, so long as the time to transmit or receive each column is nearly the same.

Non-synchronous Systems

There are very few conventional examples of non-synchronous digital systems, although most analogue modes are of this type. In these modes the receiver recognises data as and when it arrives, with no notion of synchronism. The one outstanding digital example (Morse) is typically received this way.

It does not matter in the least what speed the Morse is sent (within a range of perhaps 3:1), the data received will still be understood, whether received by ear or viewed by eye, using a lamp, a modern spectrogram or a 150-year-old clockwork Morse printer.

The only other known examples of this type belong to the MT-Hell family. In these modes the frequency of transmission, rather than the timing, describes the shape of the received character. The timing simply affects whether the received characters are short and fat or tall and thin. See Chapter 9 for more information.

References

[1] For example Clover uses a Dolph-Chebychev shape.

B

Alphabet and code reference

In this appendix
- ITA2
- Moore code
- ITA5 (ASCII)
- PSK31 varicode
- SSTV modes and VIS codes

ITA2

Letters	Figures	Code	Hex	Letters	Figures	Code	Hex
A	-	11000	18	Q	1	11101	1D
B	?	10011	13	R	4	01010	0A
C	:	01110	0E	S	'	10100	14
D		10010	12	T	5	00001	01
E	3	10000	10	U	7	11100	1C
F		10110	16	V	=	01111	0F
G		01011	0B	W	2	11001	19
H		00101	05	X	/	10111	17
I	8	01100	0C	Y	6	10101	15
J	BELL	11010	1A	Z	+	10001	11
K	(	11110	1E	CR	CR	00010	02
L	)	01001	09	LF	LF	01000	08
M	.	00111	07	LTRS	LTRS	11111	1F
N	,	00110	06	FIGS	FIGS	11011	1B
O	9	00011	03	SPACE	SPACE	00100	04
P	0	01101	0D	BLANK	BLANK	00000	00

Notes:
Unassigned FIGURES characters in the above table vary regionally.
The BLANK is rarely used. In some Amateur systems, lower case letters can be printed using BLANK to indicate a third 'shift'.
The data bits in the CODE column are transmitted in left to right order (MSB first).

Moore Code

Letters	Figures	Code	Hex	Letters	Figures	Code	Hex
A	-	1110001	71	U	7	0111001	39
B	?	0100111	27	V	=	0011110	1E
C	:	1011100	5C	W	2	1110010	72
D		1100101	65	X	/	0101110	2E
E	3	0110101	35	Y	6	1101010	6A
F		1101100	6C	Z	+	1100011	63
G		1010110	56	CR	CR	0001111	0F
H		1001011	4B	LF	LF	0011011	1B
I	8	1011001	59	LTRS	LTRS	0101101	2D
J	BELL	1110100	74	FIGS	FIGS	0110110	36
K	(	0111100	3C	SPACE	SPACE	0011101	1D
L	)	1010011	53	BLANK	BLANK	0101011	2B
M	.	1001110	4E				
N	,	1001101	4D	CTRL1	CTRL1	1010011	53
O	9	1000111	47	CTRL2	CTRL2	0101011	2B
P	0	1011010	5A	CTRL3	CTRL3	1001101	4D
Q	1	0111010	3A	IDLE β	IDLE β	1100110	66
R	4	1010101	55	IDLE α	IDLE α	1111000	78
S	'	1101001	69	SIGNAL REPETITION		0110011	33
T	5	0010111	17				

Note:
Unassigned FIGURES characters in the above table vary regionally.
The BLANK is rarely used. In some Amateur systems, lower case letters can be printed using BLANK to indicate a third 'shift'.

APPENDIX B: ALPHABET AND CODE REFERENCE

ITA5 (ASCII) - Control characters

Character	Code	Hex	Dec	Character	Code	Hex	Dec
NUL	0000000	00	00	DLE	0010000	10	16
SOH	0000001	01	01	DC1	0010001	11	17
STX	0000010	02	02	DC2	0010010	12	18
ETX	0000011	03	03	DC3	0010011	13	19
EOT	0000100	04	04	DC4	0010100	14	20
ENQ	0000101	05	05	NAK	0010101	15	21
ACK	0000110	06	06	SYN	0010110	16	22
BEL	0000111	07	07	ETB	0010111	17	23
BS	0001000	08	08	CAN	0011000	18	24
HT	0001001	09	09	EM	0011001	19	25
LF	0001010	0A	10	SUB	0011010	1A	26
VT	0001011	0B	11	ESC	0011011	1B	27
FF	0001100	0C	12	FS	0011100	1C	28
CR	0001101	0D	13	GS	0011101	1D	29
SO	0001110	0E	14	RS	0011110	1E	30
SI	0001111	0F	15	US	0011111	1F	31

ITA5 (ASCII) - Numbers, punctuation

Character	Code	Hex	Dec	Character	Code	Hex	Dec
SP	0100000	20	32	0	0110000	30	48
!	0100001	21	33	1	0110001	31	49
"	0100010	22	34	2	0110010	32	50
#	0100011	23	35	3	0110011	33	51
$	0100100	24	36	4	0110100	34	52
%	0100101	25	37	5	0110101	35	53
&	0100110	26	38	6	0110110	36	54
'	0100111	27	39	7	0110111	37	55
(	0101000	28	40	8	0111000	38	56
)	0101001	29	41	9	0111001	39	57
*	0101010	2A	42	:	0111010	3A	58
+	0101011	2B	43	;	0111011	3B	59
,	0101100	2C	44	<	0111100	3C	60
-	0101101	2D	45	=	0111101	3D	61
.	0101110	2E	46	>	0111110	3E	62
/	0101111	2F	47	?	0111111	3F	63

ITA5 (ASCII) - Upper Case

Character	Code	Hex	Dec	Character	Code	Hex	Dec
@	1000000	40	64	P	1010000	50	80
A	1000001	41	65	Q	1010001	51	81
B	1000010	42	66	R	1010010	52	82
C	1000011	43	67	S	1010011	53	83
D	1000100	44	68	T	1010100	54	84
E	1000101	45	69	U	1010101	55	85
F	1000110	46	70	V	1010110	56	86
G	1000111	47	71	W	1010111	57	87
H	1001000	48	72	X	1011000	58	88
I	1001001	49	73	Y	1011001	59	89
J	1001010	4A	74	Z	1011010	5A	90
K	1001011	4B	75	[	1011011	5B	91
L	1001100	4C	76	\	1011100	5C	92
M	1001101	4D	77	]	1011101	5D	93
N	1001110	4E	78	^	1011110	5E	94
O	1001111	4F	79	_	1011111	5F	95

ITA5 (ASCII) - Lower Case

Character	Code	Hex	Dec	Character	Code	Hex	Dec
`	1100000	60	96	p	1110000	70	112
a	1100001	61	97	q	1110001	71	113
b	1100010	62	98	r	1110010	72	114
c	1100011	63	99	s	1110011	73	115
d	1100100	64	100	t	1110100	74	116
e	1100101	65	101	u	1110101	75	117
f	1100110	66	102	v	1110110	76	118
g	1100111	67	103	w	1110111	77	119
h	1101000	68	104	x	1111000	78	120
i	1101001	69	105	y	1111001	79	121
j	1101010	6A	106	z	1111010	7A	122
k	1101011	6B	107	{	1111011	7B	123
l	1101100	6C	108			1111100	7C	124
m	1101101	6D	109	}	1111101	7D	125
n	1101110	6E	110	~	1111110	7E	126
o	1101111	6F	111	DEL	1111111	7F	127

APPENDIX B: ALPHABET AND CODE REFERENCE

Morse - Punctuation, prosigns, numbers

ASCII	Character	Code	Hex
32	SPACE		01
34	"	·—··—·	52
36	$	···—··—	C8
39	'	·————·	5E
40	(	—·——·	2D
41	)	—·——·—	6D
43	+ (AR)	·—·—·	2A
44	,	——··——	73
45	-	—····—	61
46	.	·—·—·—	6A
47	/	—··—·	29

ASCII	Character	Code	Hex
48	0	—————	3F
49	1	·————	3E
50	2	··———	3C
51	3	···——	38
52	4	····—	30
53	5	·····	20
54	6	—····	21
55	7	——···	23
56	8	———··	27
57	9	————·	2F
58	:	———···	47
59	;	—·—·—·	35
61	= (BT)	—···—	31
63	?	··——··	4C

Morse - Letters

ASCII	Character	Code	Hex
65	A	·—	06
66	B	—···	11
67	C	—·—·	15
68	D	—··	09
69	E	·	02
70	F	··—·	14
71	G	——·	08
72	H	····	10
73	I	··	04
74	J	·———	1E
75	K	—·—	0D
76	L	·—··	12
77	M	——	07
78	N	—·	05

ASCII	Character	Code	Hex
79	O	———	0F
80	P	·——·	16
81	Q	——·—	1B
82	R	·—·	0A
83	S	···	08
84	T	—	03
85	U	··—	0C
86	V	···—	18
87	W	·——	0E
88	X	—··—	19
89	Y	—·——	1D
90	Z	——··	13
95	_	··—·—	6C
	(SK)	···—·—	68

Notes
Morse alphabet and prosign usage as given in ARRL Handbook 1992, Chapter 19.
This table is intended to be representative, not exhaustive. International Morse contains many more characters – only those in common Amateur use are shown.
ASCII values are decimal.
ASCII characters not shown are not sent (ie contain HEX 01 in table).
HEX value is used to transmit Morse from microprocessor or computer as follows:
- Send bits in order right to left, 0 = dot, 1 = space, shifting byte right until byte contains 01
- Send dots as one bit time, followed by one bit-time silence
- Send dashes as three bit times, followed by one bit-time silence
- End all characters (i.e. once byte = 01) with three bit-times silence (character space). Since each dot or dash is already followed by one bit-time of silence, this means an additional two bits of silence.
- This value can be increased for 'Farnsworth' keying.
- Add two further bit times silence when SPACE is sent, for a total of five bit times (word space).

PSK31 Varicode

Character	Code	Character	Code	Character	Code
Control, upper case		L	11010111	5	101011011
NUL	1010101011	M	10111011	6	101101011
SOH	1011011011	N	11011101	7	110101101
STX	1011101101	O	10101011	8	110101011
ETX	1101110111	P	11010101	9	110110111
EOT	1011101011	Q	111011101	:	11110101
ENQ	1101011111	R	10101111	;	110111101
ACK	1011101111	S	1101111	<	111101101
BEL	1011111101	T	1101101	=	1010101
BS	1011111111	U	101010111	>	111010111
HT	11101111	V	110110101	?	1010101111
LF	11101	W	101011101	`	1011011111
VT	1101101111	X	101110101	a	1011
FF	1011011101	Y	101111011	b	1011111
CR	11111	Z	1010101101	c	101111
SO	1101110101	[	111110111	d	101101
SI	1110101011	\	111101111	e	11
DLE	1011110111	]	111111011	f	111101
DC1	1011110101	^	1010111111	g	1011011
DC2	1110101101	_	101101101	h	101011
DC3	1110101111			i	1101
DC4	1101011011	**Punctuation, numbers,**		j	111101011
NAK	1101101011	**lower case**		k	10111111
SYN	1101101101	SPACE	1	l	11011
ETB	1101010111	!	111111111	m	111011
CAN	1101111011	"	101011111	n	1111
EM	1101111101	#	111110101	o	111
SUB	1110110111	$	111011011		
ESC	1101010101	0%	1011010101	p	111111
FS	1101011101	&	1010111011	q	110111111
GS	1110111011	'	101111111	r	10101
RS	1011111011	(	11111011	s	10111
US	1101111111	)	11110111	t	101
@	1010111101	*	101101111	u	110111
A	1111101	+	111011111	v	1111011
B	11101011	'	1110101	w	1101011
C	10101101	-	110101	x	11011111
D	10110101	0	1010111	y	1011101
E	1110111	/	110101111	z	111010101
F	11011011			{	1010110111
G	11111101	0	10110111			110111011
H	101010101	1	10111101	}	1010110101
I	1111111	2	11101101	~	1011010111
J	111111101	3	11111111	DEL (127)	1110110101
K	101111101	4	101110111		

Notes
Codes are transmitted left to right. '0' represents PSK phase reversal, '1' represents no phase change.
At least two '0's are inserted between characters.

APPENDIX B: ALPHABET AND CODE REFERENCE

SSTV Modes and VIS Codes

Family	Mode	Colour	Secs	Lines	VIS	Notes
AVT	AVT24	RGB	24	120	4n	1,7
	AVT90	RGB	90	240	44+n	1,7
	AVT94	RGB	94	200	48+n	1,7
	AVT188	RGB	188	400	4C+n	1
	AVT125	BW	125	400	5n	1,7
Martin	M1	RGB	114	240	2C	2
	M2	RGB	58	240	28	2
	M3	RGB	57	120	24	3
	M4	RGB	29	120	20	3
	HQ1	YC	90	240	29	
	HQ2	YC	112	240	2A	
Pasokon	P3	RGB	203	496	71	4
Hires	P5	RGB	305	496	72	4
	P7	RGB	406	496	73	4
Acorn	PD90	YC	90	240	63	
	PD120	YC	126	480	53	
	PD160	YC	161	384	62	
	PD180	YC	187	480	60	
	PD240	YC	248	480	61	
Robot	Robot8	BW	8	120	01-03	5,6
	Robot12	BW	12	120	00	8
	Robot12	YC	12	120	05-07	6
	Robot24	BW	24	240	04	8
	Robot24	YC	24	120	09-0B	6
	Robot36	BW/YC	36	240	08	6
	Robot72	YC	72	240	0C	
Scottie	S1	RGB	110	240	3C	2
	S2	RGB	71	240	38	2
	S3	RGB	55	120	34	3
	S4	RGB	36	120	30	3
	DX	RGB	269	240	4C	2
Wrasse	SC1/24	RGB	24	120	10	3
	SC1/48	RGB	48	240	14/18	2
	SC1/96	RGB	96	240	1C	2
	SC2/30	RGB	30	128	33	
	SC2/60	RGB	60	256	3B	
	SC2/120	RGB	120	256	3F	
	SC2/180	RGb	180	256	37	

Notes
This list is not intended to be exhaustive or authoritative. Several modes are known to use the same VIS codes. New modes continue to be developed. Modes M1 and S1 are by far the most common.

1. 5 sec digital header, no sync
2. First 16 lines grey scale, 240 usable lines
3. First 8 lines grey scale, 120 usable lines
4. First 16 lines grey scale, 480 usable lines
5. Close to the original BW standard
6. Red/Green/Blue components sent separately
7. Subversions have different VIS codes n = 0 – 3
8. Green component sent to simulate BW

Appendix C

Software & information sources

In this appendix
- Digital mode software
- Multi-mode programs
- Resources

SOFTWARE FOR amateur digital modes is generally well documented, so there is no need to describe how to operate the programs here. The trickiest part is setting up the sound card to operate with the software, and this is covered in depth in Appendix D. There is such a wide variety of digital mode software that it would not be possible to review more than a small proportion of it. This chapter includes a brief description of popular and representative software, along with references to where to find the software. Most of the software is available from several sources, but if there is a 'home page' for the software, it is more likely to have the latest version. Internet references are notoriously ephemeral, so if there is a problem finding the software, use a search engine. The quality, performance and documentation of most Amateur sound-card software is generally of a remarkably high standard, and yet almost all the software is free!

Software is offered at several levels, starting with experimental (use at your own risk, some expertise required), through freeware, completely developed but also completely free to use (often for Amateur use only), then shareware, which has restricted free use (maybe some features are disabled, or the time period for free use is limited), and is then registered for full use, usually at a modest price. Finally commercial software, which you pay for before you use it, although often there are demo versions available as well. Commercial software is generally the most expensive, but usually also has the most features and is better supported. User support for freeware is generally to be found on e-mail reflectors (specialist mailing lists). Shareware is generally supported by the authors, but only for registered users.

The Internet is very rich in papers, articles and web sites dedicated to the digital modes. This Chapter contains references to many of these, but of course is by no means exhaustive. Where there are several useful subjects, papers or articles on one site, the main site reference is given, and the reader will need to look around to find the nuggets.

Digital Mode Software

Fax & Satellites – WXSAT
This software, by Christian Bock, gives high-resolution satellite images, and is pre-configured for a wide range of satellite formats. It also works very well with HF FAX transmissions such as weather maps. The software is free and works best with a high-resolution monitor. Pictures can be saved in standard BMP image format, although the files can be very large. The software can be set to start automatically on a schedule, including the correct mode settings, and also recognises transmitted start and stop signals if the signal is clear enough.

Hellschreiber – IZ8BLY Hell
The definitive Hell software. Nino IZ8BLY has been at the forefront of modern fuzzy-mode development. His software is free, gives very high performance, is easy to use, and offers all the popular Hell modes, and some others as well. The software interacts with programs for other modes, and includes logging. This program can also be used for automated Hell beacons. See the IZ8BLY website for all Nino's software [1].

MFSK16 – Stream
The first software to offer MFSK16, and still the only software with MFSK8 and PSK63F. This is the reference program to which others are compared, and is free. It includes a very clear tuning display, and is an easy to use program with the same 'look and feel' as other IZ8BLY software. The software started life as a 'test platform' used by the development team for testing new modes and ideas. It also operates PSK31.

Morse – CWGet/CWType
This interesting software by Sergei UA9OSV provides the best computer copy of Morse of any of the available software, free or otherwise. This software is shareware, but the receiving program CWGET demonstrates excellent performance even before registering. The receiver program features auto-tuning, so can track a QSO when stations are on different frequencies. It also tracks varying signal levels moderately well (always the most difficult task for computer Morse) and tracks speed changes and poor weighting well. Performance is excellent from 10 WPM to 40 WPM if signals are strong and stable. The transmit program uses direct keying of the rig in CW mode (not via the sound card). This allows the two programs to be used together for full duplex, testing or QRK (full break-in) [2].

MT63 – IZ8BLY MT63
The software which revolutionised this mode, introducing MT63 to a wide audience. Excellent tuning and high performance with very robust sync. IZ8BLY MT63 is free, has all the popular speed options, logging, auto-reporting, CWID on CQ and is the only software with the secondary channel ID message on transmit and receive.

PSK31 – Digipan
Excellent free software for PSK31 and QPSK31. Digipan has a good tuning display suited to crystal controlled transceivers, can receive on two frequencies at once, and transmit on either frequency. Software is available from many places, but check the Digipan website [3]. There is a shareware program from the same stable that includes many other modes. See MIXW 2 in the multi-mode section.

RTTY – MMTTY
MMTTY is the definitive RTTY program for PC sound card. The performance is excellent, and it is slick and simple to use. It has a wide range of Amateur and commercial settings to choose from, and excellent tuning aids. The filters used are very high performance, and several different choices are offered for the user to experiment with. See the MMTTY web site [4].

SSTV – MMSSTV
Very easy to use, MMSSTV also performs very well. The software is free, has a 'history' feature that allows the user to retransmit received pictures either full size or as an insert, has a range of flexible text addition options, and can use just about any picture format by drag and drop or cut and paste. A large library of pictures can be stored ready to transmit at a moment's notice, and of course pictures for transmission can be prepared while receiving. MMSSTV has a wide range of modes that are detected automatically using the VIS. In most cases the correct mode can be detected even once the picture has started. See the MMSSTV web site [5].

Multi-mode Programs

MIXW 2
This is a really spectacular program, which works extremely well and is widely used. It is shareware, but not expensive to register. MIXW 2 operates all the popular modes: RTTY, PSK31, AmTOR FEC, MFSK, Packet, Hellschreiber, Throb, SSTV and MT63, and will also receive FAX and PacTOR. Not only does MIXW include logging, but will also log for modes it does not operate, such as SSB, FM and CW [6].

TrueTTY
Not quite as versatile as MIXW 2, TrueTTY operates RTTY, PSK31, AmTOR FEC and Packet. The software is by UA9OSV and has similar clever auto-tuning and the excellent filters of his CWGET software. TrueTTY includes logging capability and is shareware [7].

HamScope
This program started life as an experimental platform, and still has some interesting experimental features, such as selectable error correction options. It allows the user to experiment with and evaluate a wide range of ECC techniques (most of which work poorly on HF, but using this software is a great way to prove the point!). HamScope runs RTTY, using the MMTTY 'engine', MFSK16, and PSK31. The software is by Glen Hansen KD5HIO, and is freeware [8]. Be aware that most of the ECC options offered are non-standard, and

there will be no way for other Amateurs to receive the transmissions except by prior arrangement and agreement of protocols.

IZ8BLY Suite
Multi-mode capability can be achieved by combining IZ8BLY Hellschreiber, MT63 and MFSK16 software, which can all be operated together by being selected one from another. Thus a wide range of modes is available. All the Hellschreiber modes, MT63, MFSK16, MFSK8, PSK31, PSK63F, PSK125F and PSK250F can be selected. This software is free, and the programs have the same 'look and feel'.

References
[1] http://iz8bly.sysonline.it/
[2] See http://www.dxsoft.com/
[3] http://members.home.com/hteller/digipan/
[4] http://www.qsl.net/mmhamsoft/mmtty/index.html
[5] http://www.qsl.net/mmhamsoft/mmsstv/index.htm
[6] See http://tav.kiev.ua/~nick/mixw2/
[7] See http://www.dxsoft.com/

Resources

Amateur Software Sites

EA3QP Amateur software	http://personal3.iddeo.es/ea3qp/soft1.html
NL9222 Visual modes	http://home.wanadoo.nl/nl9222/software.htm
NL9222 Digital modes	http://home.wanadoo.nl/nl9222/digisoft.htm
DL9QJ Soundcard software	http://www.muenster.de/~welp/sb.htm
Linux Amateur software	http://radio.linux.org.au/

Mode Reference Sites

FAX	http://www.hffax.de
PSK31	http://www.kender.es/~edu/psk31.html
	http://www.psk31.com/
Fuzzy modes (Hell)	http://www.qsl.net/zl1bpu
	http://sharon.esrac.ele.tue.nl/zl1bpu/index.html
MFSK16	http://www.qsl.net/zl1bpu/MFSK/
MT63	http://www.qsl.net/zl1bpu/MT63/MT63.html

Other Software

ZL2AKM (BTL for DOS)	http://www.geocities.com/SiliconValley/Heights/4477/
MSCAN SSTV	http://www.mscan.com/
MIXW	http://tav.kiev.ua/~nick/mixw/mixw.htm
Clover	http://www.halcomm.com/
PacTOR	http://www.scs-ptc.com/

continued overleaf

continued from previous page

Throb etc	http://www.lsear.freeserve.co.uk/
FSK441	http://pulsar.princeton.edu/~joe/K1JT
VE2IQ PSK modes	http://cafe.rapidus.net/bill1/bbs.htm
ALE, STANAG 4285	http://www.chbrain.dircon.co.uk/
Chirpsounding	http://www.qsl.net/zl1bpu/chirp/chirps.html
	http://www.asenior48.freeserve.co.uk/chirpview.html

Hardware

Rigblaster	http://www.westmountainradio.com
Small Wonder Labs	http://www.smallwonderlabs.com/index.html
NJQRP kits	http://www.njqrp.org/
PIC based PSK31 RX	http://www.users.bigpond.com/gzimmer

History

Hellschreiber	http://www.qsl.net/zl1bpu/FUZZY/History/History.html
RTTY history	http://www.samhallas.co.uk/telhist1/telehist.htm
Five-unit codes	http://www.nadcomm.com/fiveunit/fiveunits.htm
SSTV	http://www.darc.de/distrikte/g/T_ATV/sstv-history.htm

Technical Information

Tutorial on Viterbi decoding	http://pweb.netcom.com/~chip.f/Viterbi.html
KA9Q FEC references	http://people.qualcomm.com/karn/code/fec/
Ionospheric simulation	http://www.peak.org/~forrerj
Error coding tutorial	http://www.ittc.ukans.edu/~paden/
reference/guides/ECC/introduction.html	
Hamming codes	http://www.engelschall.com/u/sb/hamming/
PSK31 Fundamentals	http://aintel.bi.ehu.es/psk31theory.html

Clubs and Groups

BARTG British Amateur Radio Teleprinter Group	http://www.bartg.demon.co.uk/index.htm
BATC British Amateur Television Group	http://www.batc.org.uk/index.htm
RIG Remote Imaging Group	http://www.rig.org.uk
RNARS Royal Navy Amateur Radio Society	http://www.rnars.org.uk/index.html
RSARS Royal Signals Amateur Radio Society	http://www.rsars.org.uk
RSGB Radio Society of Great Britain	http://www.rsgb.org/
RSGB Data Communications Committee	http://www.dcc.rsgb.org/
TAPR Tucson Amateur Packet Radio Inc	http://www.tapr.org/

D

The PC sound card

In this appendix
- Inside the sound card
- Sound card setup for digital modes
- Receiving setup
- Transmitting setup

IN ORDER TO operate digital modes using a sound card, it is not important to know how the sound card works, but it is very important to know how to set it up. In order to make this process easy to understand, it is however, necessary to have a rudimentary knowledge of what constitutes a sound card or sound interface, and how it is controlled. This Appendix includes a simple guide to what is inside a sound card.

The control of the sound card is largely performed by software provided by the Windows operating system, rather than the individual digital-mode programs. This means that it is possible to explain how this software interacts with the computer sound card, and explain the setup procedure for the sound card, independent of the actual application. We can be confident that the procedure will then be the same for all programs.

Inside the Sound Card

The most popular and most widely-supported sound card is the 'Soundblaster' type (see **Fig D.1**) [1]. Many varieties of Soundblaster cards have been available over the years, and all but the very early ones are quite appropriate for digital modes (just avoid the oldest eight-bit cards). There are even more clones of the Soundblaster type, which function the same even if they are not electrically identical. The secret is in the driver software, which makes the sound card appear the same as other cards or sound interfaces to the operating system.

It is best to avoid other types of sound cards. While there are other cards with outstanding features, such as high-fidelity sound, unless they have Soundblaster functionality they may not have the features required for the operating software. The good news is that Soundblaster-type sound cards are reasonably cheap, and it is also practical to have more than one sound card, even of different types, installed at the same time.

175

Fig D.1: A typical PC sound card

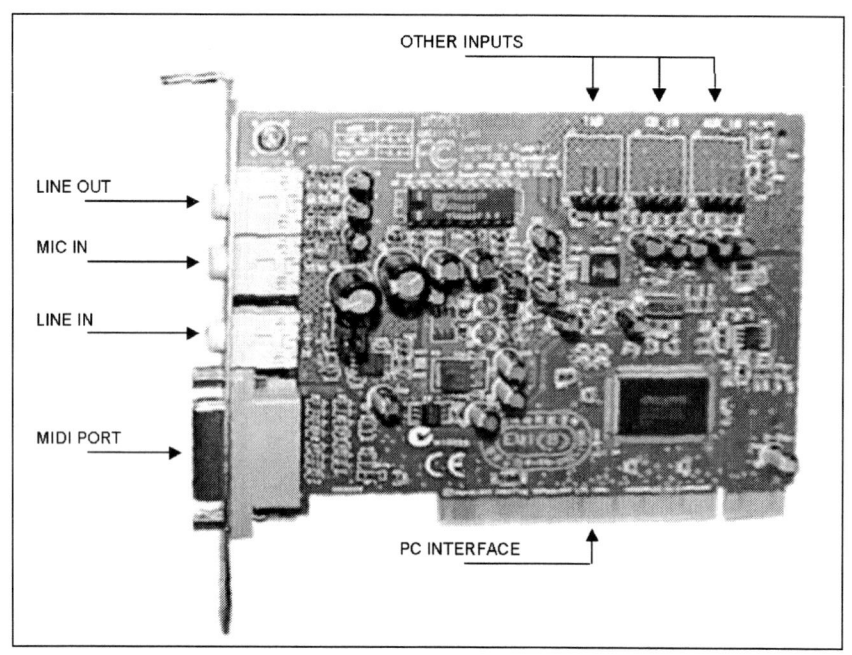

The transparency to the operating system provided by the driver software is the secret to reliable operation with a wide variety of different cards with a wide variety of different programs. The driver software provided by the manufacturer ensures that the interface to Windows is consistent, so any Windows application wishing to use the sound card will have a consistent interface – in theory. Generally this approach works well, and is a significant advance from past systems where each software manufacturer was forced to develop drivers for every hardware configuration they wished to support, and this was of course very limiting.

The consistent interface is provided by a program called SNDVOL32, and the standard Soundblaster version of SNDVOL32 comes with Windows and is used by the vast majority of sound cards. Setting up the sound interface for digital modes using this program will be described later in this Appendix.

Sound cards are invariably stereo devices these days, but for Amateur purposes the two channels are usually operated together, so mono connections will usually suffice, and the operator can generally forget about the stereo nature of the system. The cables described in Chapter 6 are mono cables.

The sound card contains several functional blocks (see **Fig D.2**). There is usually a music synthesiser, which uses electronics to generate sounds. This is an FM synthesiser capable of generating nine or more different and independently controlled tones, plus noise and percussion effects. This section is not often used for digital modes. The synthesiser is controlled by PC data or by special music commands (called the MIDI system). Higher-performance sound cards also have a wavetable synthesiser, which can generate musical or other synthetic sounds by replicating them from a large table of data loaded by the computer. This system is also MIDI controlled. Although widely used for very realistic music, this section is not used for digital modes. The section that is of most interest to Amateurs is the sound sampling system, sometimes called the Wave system, since it can record and play waveforms like a digital tape recorder.

Central to the sound card and its operation is a mixer, or rather two mixers, one for input ('receive' or 'record') and one for output ('transmit' or 'playback'). These mixers have many inputs and are completely controlled by the sound-card driver software. On modern cards even the mixers are completely digital.

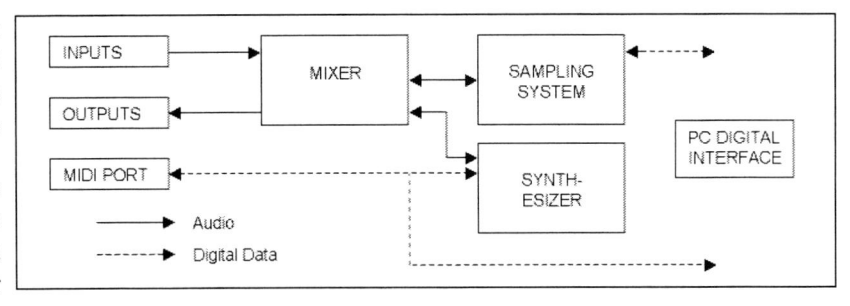

Fig D.2: Sound-card block diagram

The Sampling System

The sampling system is bi-directional. The input section measures the sound-card audio input signal from the record mixer rather as a digital voltmeter does, and reports the voltage measured to the PC. It operates very quickly – at speeds up to 44,000 samples per second. This is the sampling rate, which has a direct effect on performance. The *sampling rate* must also be stable and accurate.

For Amateur purposes, the sound card must be at least 16 bit, ie the resolution of the sampling system must be good enough to resolve 2^{16} or 65536 different voltage levels. The resolution affects the level of distortion. Low resolution adds noise to the receiving process and causes the transmitted signal to be less clean than is otherwise possible.

The output sampling section converts digital numbers into voltages, so data sent to the sampling system at the sampling rate will be converted into real sounds which are sent to the playback mixer and on to the sound-card audio outputs.

In order for sound to be replicated accurately, either as digital samples of input sounds, or digital representations of output sounds, sampling theory tells us that the sampling rate (data rate) must be much higher than the signal audio frequency. For most radio applications with limited audio range (say up to 2.4kHz), and for modest fidelity, sampling rates of 8000Hz or 11025Hz are used, ensuring at least three or four samples per cycle. As the sampling rate is reduced and approaches two samples per cycle, 'alias' signals, just like receiver mixer images, appear on the signal.

One of the important functions of the receiving sound system is to ensure that high-frequency sounds from the radio (say above 3kHz) are kept out of the sound system. If these sounds are allowed in with the wanted signal, they will cause audible aliases that will mix with the wanted sounds, degrading the performance. Controlling these unwanted signals is best done in the receiver using a narrow filter, but some radio software also does this using a 'decimation filter', where the signal is sampled faster than necessary, and then averaged to remove the high frequencies, rather like a software low-pass filter. This technique is very effective, and is also very fast so uses little computer power.

The sampling system must also operate at an accurate and steady speed, and therefore requires the PC to receive and use the data in a timely manner. On transmit the PC needs to provide the sound card with data when it needs it. This can be difficult to achieve with Windows software, since the speed of the operating program is not always constant. Fortunately the interface between the sound card and the computer is buffered in both directions (rather like the keyboard buffer in Amateur software), and the sound card and PC communicate with each other in order to manage the data correctly. This is all provided in the driver software, so the software-application writer, and especially the user, do not need to be concerned with how this happens.

Software for Amateur digital modes operates at very high speed, processing digital samples of the signals at the sampling rate. This involves many repeated mathematical computations, almost all at the sampling rate. Therefore the computer needs to have considerable processing power in order to complete all the calculations in time for the next set of calculations just a few microseconds later. The processor should operate at a clock speed of least 100MHz, and must include a maths co-processor in order to handle the mathematics at the required speed.

Inputs, Outputs and Mixers

The sound card has many audio inputs – at least a line input, a microphone input and a CDROM audio input (for playing music). These inputs are managed and selected by a mixer built into the sound card, rather like a recording studio mixer, but controlled by software. This mixer is the 'record mixer', and sends data to the input of the software application. In our case, that's the digital-mode 'receiver' input.

Similarly, the sound card has many output sources – all the input sources can be selected for output, as well as the synthesiser outputs (the wave sampling system, the MIDI synthesiser and the wavetable synthesiser). These outputs are selected and controlled by another mixer, called the 'volume control' (it should be called the playback mixer, and the name does tend to cause confusion).

There can be other inputs and outputs from modems, TV and radio cards, Internet phone systems, system sounds and so on, and these can cause confusion when trying to set up for digital modes. It is best to keep these disabled and switched off, otherwise the software may not work, or unpredictable results may occur.

Fig D.3: The Record mixer

The mixers are central to the correct setting and operation of Amateur digital modes, so their use will be explained very carefully. Study the block diagrams shown in **Fig D.3** and **Fig D.4**.

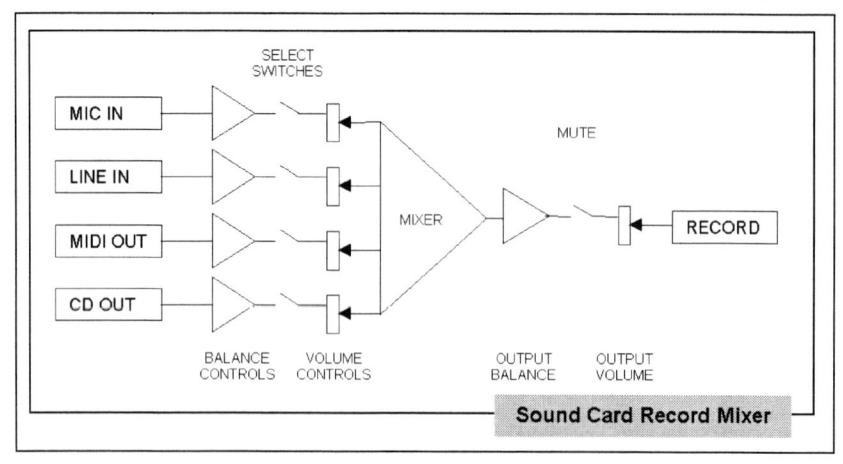

Record Mixer

The inputs shown in Fig D.3 are representative. Your computer may

APPENDIX D: THE PC SOUND CARD

have several more. These can be added or disabled, so some may not appear. A description will be given later about how to add and remove these inputs from the mixer.

Simply ignore the Balance controls, and leave them set to midway. If they are inadvertently left hard one way or the other, it may be very difficult to work out what is wrong!

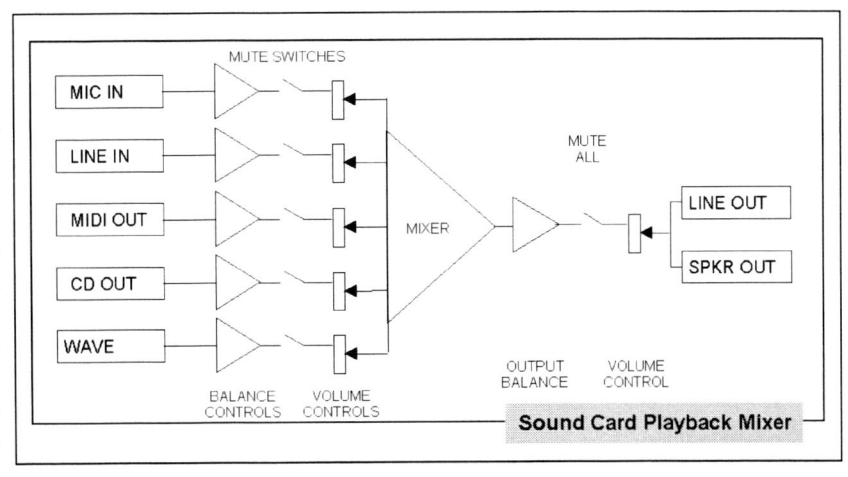

Fig D.4: The Playback mixer

The Select switches allow the active inputs to be selected. On some computers only one may be selected at a time, but on most computers more than one can be selected. For digital modes, preferably select the Line In and turn off the rest. When the input is selected, a tick shows under that device on the mixer controls. Leaving the required input unselected is one of the most common causes of non-operation of digital-mode software.

The Volume controls act as would be expected. They adjust the input level to the Amateur software from the input source, typically the receiver. With typical speaker-level audio input, half way would be a good place to start. Take care not to leave the Volume for the selected input fully down, as it might prove difficult to find the cause when the software does not operate.

The actual mixer is next. All active inputs will have their sounds blended into the audio sent to the software. Obviously, if there is audio coming in via these inputs the effect is exactly the same as interfering signals coming in from the receiver. Be especially careful to disable the Mic In, or sounds within the shack will interfere with reception.

The output of the mixer is a single (stereo) sound channel. This is the audio signal sent to the sampling system, and therefore to the application software record input (the record section or receiver section, depending on the application). The mute control disables the output, and finally the Output or Recording Volume sets the actual audio level to the software. Half way is a good starting setting for this control as well. Sometimes the sound card does not have the hardware for the final Mute or Recording Volume, and the corresponding controls will not work (they will be 'greyed out' on the control panel).

Later in this Appendix, the adjustment of these controls will refer to the corresponding Recording Control panel. When making these adjustments, bear in mind the block diagram in Fig D.2, in order to better understand what is going on.

Playback Mixer

As with the record controls, the inputs shown in Fig D.4 are representative. Your computer may have several others, such as advanced wavetable synthesisers, TV and radio audio, Internet phone etc. Notice that some of the inputs are the same as the Record inputs. These are inputs from the 'outside world' as far as the sound card is concerned – microphone, line input and CD audio. The other two are internal sound-card sources, the MIDI synthesiser and the Wave output from the sampling-system playback. This last input is the one of interest for digital modes. This input (to the mixer) is the output from the sampling system which makes sounds generated in the Amateur radio software – the transmit signal.

All the inputs described can be added or disabled, and some may not appear. It is unlikely that the necessary Wave output will be missing, but a description will be given later showing how to add and remove these inputs. As with the record controls, simply ignore the Balance controls, and leave them set to mid way.

The Mute switches allow the active inputs to be disabled. On most computers, more than one can be active at a time. For digital modes, preferably select the Wave input and turn off the rest. For example, you must disable the microphone input or your keyboard clatter and muttered noises will be sent over the air! When an input is muted (disabled), a tick shows under that device on the mixer controls. Be aware that this mute operation acts in a reverse manner to the Record controls.

The Volume controls adjust the output level from the Amateur software to the transmitter or PC speakers. The transmit interface cable should be designed so that with normal transceiver microphone audio level, the Wave volume control will be about half way up. This is not a critical adjustment.

All active inputs are blended in the mixer to produce the audio sent to the sound card output (and the transmitter). Obviously it's not a good idea to have other sources such as the MIDI synthesiser or CD audio enabled while operating Amateur modes!

The output of the mixer is a single (stereo) sound channel. The mute control 'Mute All' disables the output, and finally the Output or Recording Volume sets the actual audio level to the software. Half way is a good starting setting for this control as well. This control often has more range than the individual channel controls, so this is the best control to use to set the transmitter drive. Also, be aware that this control is the same control that appears in the system tray (bottom right corner of the Windows screen) as the volume control. Before transmitting, check the setting, as it may have been adjusted for other applications. Of course this is very convenient as the system tray control does the trick without opening the control panel (**Fig D.5**).

The adjustment of these controls for digital modes will be covered in detail later in the next section of this Appendix. When making these adjustments, refer back to the block diagram in Fig D.2.

Sound Card Setup for Digital Modes

Because the sound-card settings tend to interact, and are likely to be different from how the sound card is set up for other (non-Amateur) applications, it is best to make the adjustments in the order suggested, and note the settings for future reference.

A very simple way to do this is to print or save a 'snapshot' of the recording and playback control panels. If the adjustments are subsequently changed, it will then be the work of a moment to restore normal digital-mode operation.

(Left) Fig D.5: The Win98 System Tray

APPENDIX D: THE PC SOUND CARD

The setup that follows is described for the IZ8BLY Hellschreiber software. This software has been chosen for several reasons. First, Hell is an easy mode to get started with, is not radio-sideband dependent, and is reasonably easy to set up. Secondly, it is easy to tell from the software when the receive and transmit audio is working and correctly adjusted. Finally, this software has buttons on the tool bar for the sound-card controls, making the setup easy for those unfamiliar with the process or not familiar with the finer techniques of PC control.

(Right) Fig D.6: Right clicking on the Speaker in the System Tray

Before proceeding further with the setup, download and install the IZ8BLY Hellschreiber software [2]. Once you have the Hellschreiber software operating correctly, install some other modes and check them out. The correct settings will probably be exactly the same.

Some Amateur mode software does not include sound-card controls. For these you must find the controls manually. These instructions apply to Windows 95, 98 and 2000, and to the standard Soundblaster control panels. Other operating systems and sound-card controls may differ slightly. Here is the procedure.

Recording Controls

This is the harder of the two control panels to find. For software that does not include links to these controls, there are two methods, and the easiest is as follows:

(1) From the system tray (bottom right corner of Windows screen), *right click* on the Volume Control icon (ie point to the Speaker icon with the mouse, and click the right mouse button) (**Fig D.6**).

(2) *Left click* on **Open Volume Controls**. This will open the Playback control panel (not the one you want!) (**Fig D.7**).

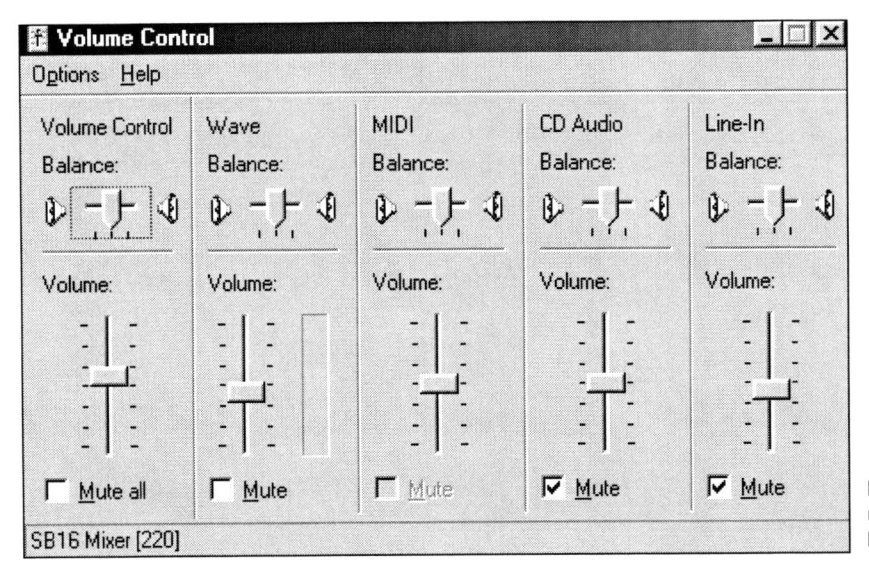

Fig D.7: The Volume Control (Playback) panel

181

Fig D.8: Selecting 'Options/Properties'

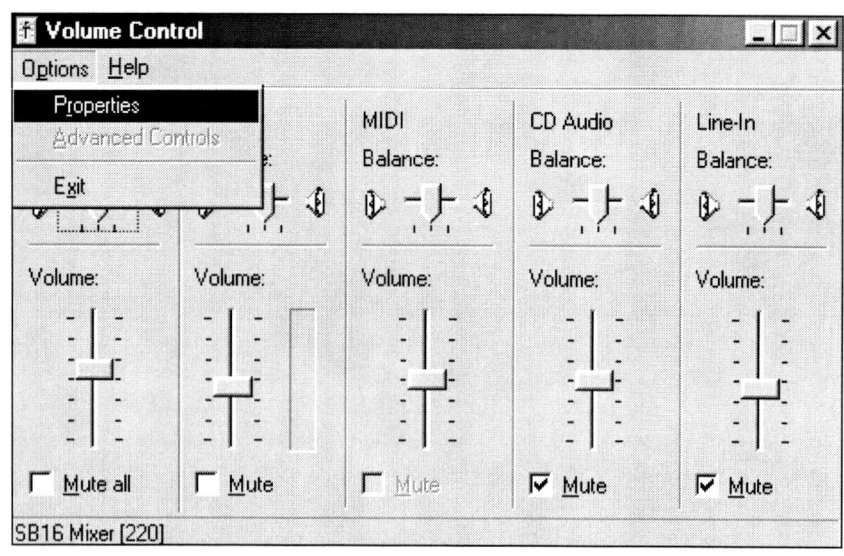

(3) On this panel, *left click* on the menu item Options and select Properties. (In Windows jargon, this is described as Options/Properties) (**Fig D.8**).

(4) *Left click* on the word **Recording**, and note the 'radio button' turns on next to the word you clicked on.

The lower panel shows the recording devices that are active – more about that later (**Fig D.9**).

Fig D.9: Selecting the 'Recording' radio button

(5) *Left click* on the **OK** button, and the recording panel will magically appear! (**Fig D.10**).

Yes, that's the easiest method! You now understand the reason for the Tool Bar buttons in the software. There is one further problem.

In Windows 95 there is a 'feature' that prevents the application software from correctly calling up the Record control. Instead, the Playback control (labelled Volume Control) is called up. This leads to considerable confusion among beginners. When using Windows 95 and this happens, force the Recording control to appear by following the procedure above from Step 3.

The other method is to open the Windows Control Panel (**Start/Settings/Control Panel**), and *double click* (two left mouse clicks while pointing to the icon) on **Multimedia**. Then *left click* on the microphone icon about half

182

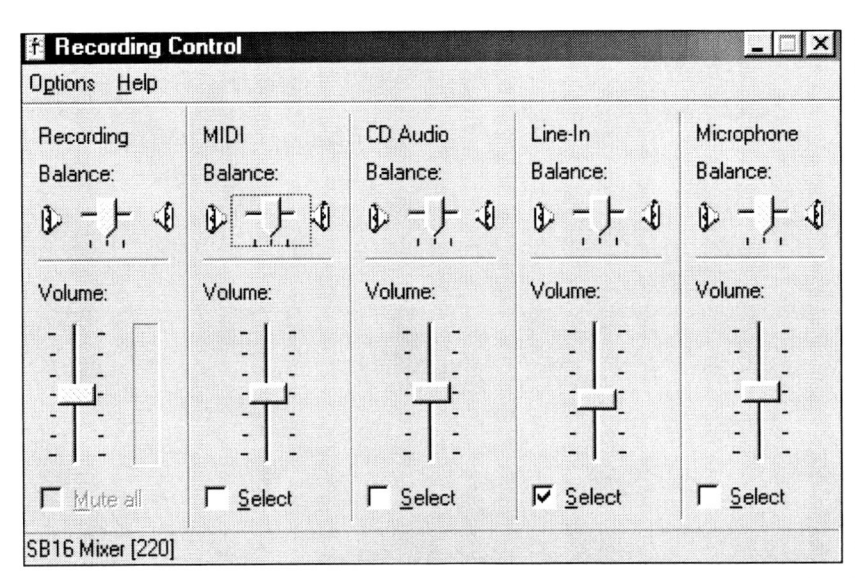

APPENDIX D: THE PC SOUND CARD

(Left) Fig D.10: The Recording Mixer panel

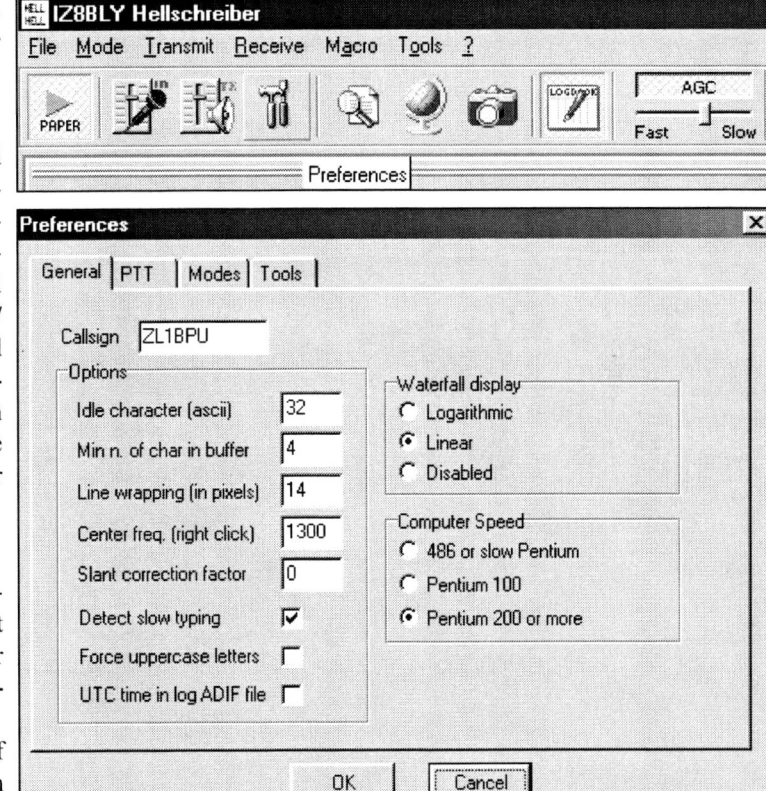

(Below) Fig D.11: The IZ8BLY tool bar

way down the panel, under the word **Recording**.

Playback Controls
Proceed as described under Recording Controls, up to Step 2. Alternatively, open the Windows Control Panel (**Start/Settings/ Control Panel**), and *double click* on **Multimedia**. Then *left click* on the speaker icon near the top of the panel, under the word **Playback**.

Receiving Setup
From the IZ8BLY software 'Tool Bar', select the icon with the hammer and chisel on it (**Preferences**) (**Fig D.11**).

On the **General** tab of the panel that appears, in the area labelled **Waterfall display**, click on **Linear**. Then click on the **OK** button. This sets the waterfall into linear mode, with black signals on a white background (**Fig D.12**).

From the IZ8BLY Toolbar, now select **Set input volume**, the icon with a slider and a microphone (second icon from the left in Fig D.12) to open the Windows Recording Control Mixer panel. The Recording Controls panel will

Fig D.12: The Preferences panel

183

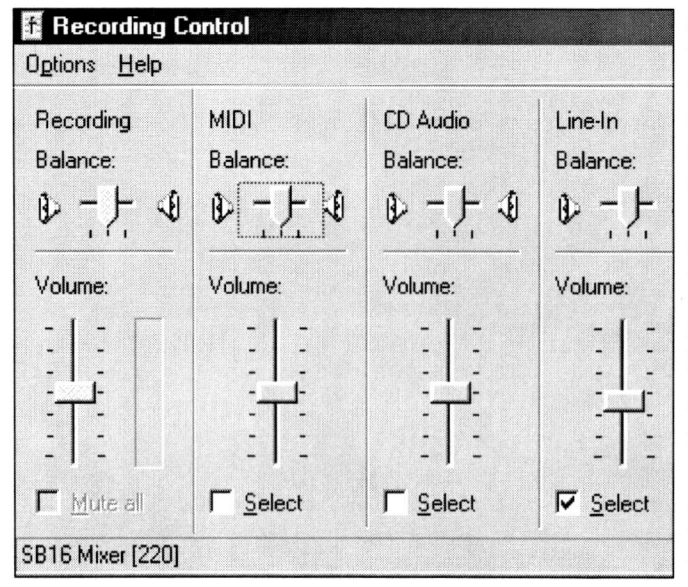

appear on top of the IZ8BLY software. (If you are using Windows 95, the incorrect panel will appear, so on the panel that does appear, select **Options/Properties**, and click on the **Recording** button).

On the Recording Control panel, select **Options/Properties**. Check that the Recording radio button is present. Below the radio buttons, you will see a window containing the currently available recording controls. Make sure the input you need (**Line Input** or **Mic Input**) is ticked (checked) to make it available on the Recording Controls. If it is not ticked, *left click* in the box to make the tick appear. These controls set the level of signals from the outside world that go to any recording software, such as the IZ8BLY Hellscreiber receiving display. Any inputs you don't ever plan to use can be deselected (click on the tick to make it disappear). *Left click* on the **OK** button (**Fig D.13**).

Fig D.13: Selecting the Record input controls

The Recording Control panel is divided into sections, each labelled with an input source name above. Below are the balance, volume and selector controls for that input. **Select** the check box for the input you are using (*left click* on the word **Select**), and make sure the rest are not selected (**Fig D.14**).

Set the receiver audio to a normal listening level, and make sure that both the radio speaker and the sound-card audio cable are receiving audio. Check that the sound-card cable is plugged into the sound-card input you have just selected.

Fig D.14: The recording controls

Tune the receiver to a clear spot on a noisy band, and watch the IZ8BLY screen for background noise. Adjust the **Line Input Volume** slider (or **Mic Input** if you are using that) to give a modest amount of speckled grey noise on the screen. If nothing appears, recheck the receiver-cable connections and that the correct input is selected.

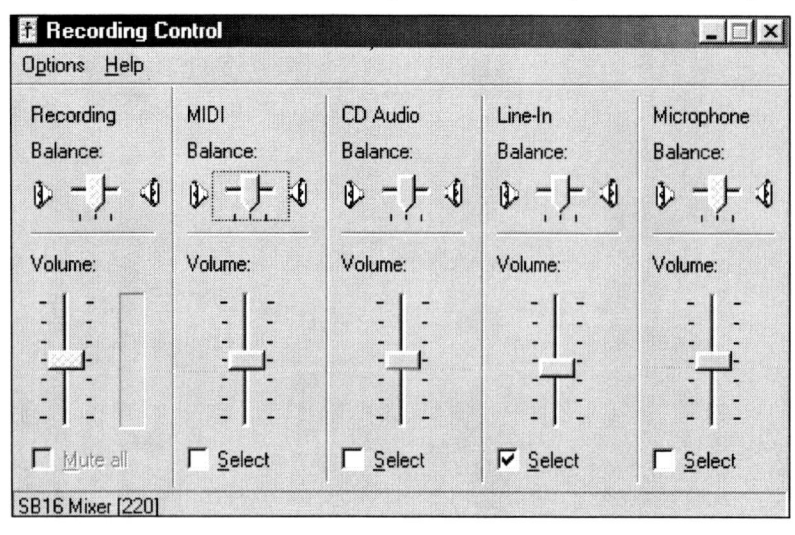

Hint: As a digression, if you have trouble working out if there is audio coming into the sound card from the receiver, open the **Playback**

APPENDIX D: THE PC SOUND CARD

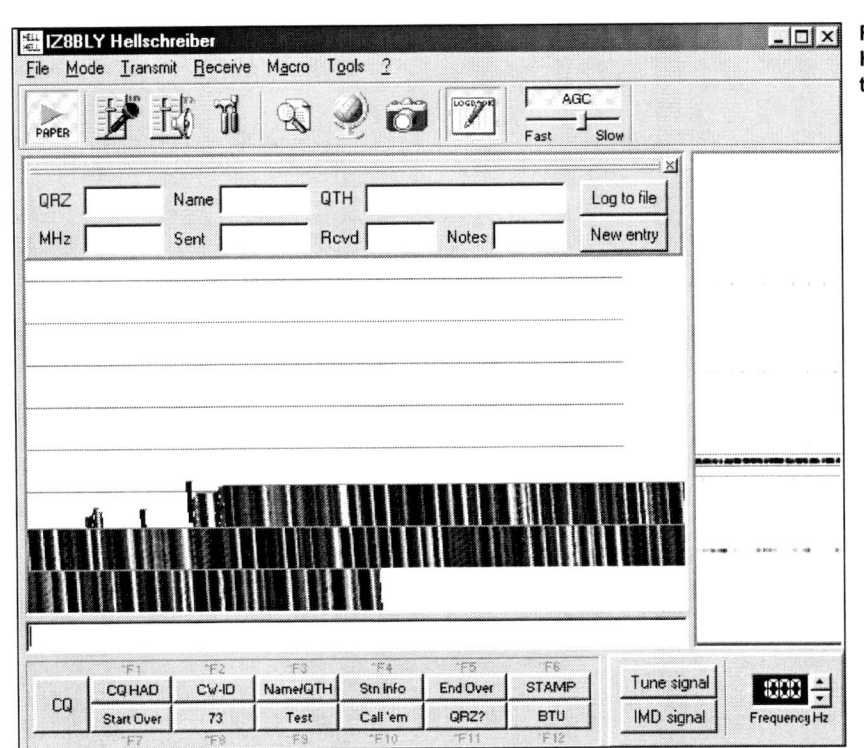

Fig D.15: IZ8BLY Hellschreiber tuned to a carrier

controls (labelled 'Volume Controls'), select **Line Input** (or **Mic Input**) as an output, connect up computer speakers to the Line Out or Speaker Out sound-card socket and check that you can hear the receiver signal from the computer speakers.

Now tune in a carrier or Morse signal, on the receiver, adjusting it to give about a tone of about 1kHz (for example, tune in a frequency-standard on USB by dialling up 9.999 MHz). You should see a fuzzy line on the waterfall display (right side), but don't worry if you don't. Below the waterfall display is a small frequency display. Change the setting to read '1000'. There are three coloured lines on the waterfall display (two blue and one red), which will move as you change the frequency setting. A black, slightly fuzzy line should be seen on the waterfall near to the tuning lines, and a broad black bar should appear on the main display. Adjust the **Line Input Volume** slider (or **Mic Input**) up until the signal display on the waterfall shows flashes of red instead of black (this indicates overload) and then drop the level back down just until there is no sign of red. IZ8BLY Hellschreiber tuned to a carrier is to be seen in **Fig D.15**.

Don't worry if you do not see the change to red, as long as the signal can be seen. In this case adjust the volume slider so that you clearly see a black line. Experience has shown that sound-card sensitivity varies, and it is sometimes not possible to see the waterfall display. If your PC is very slow, the waterfall may not work anyway. Record all the settings on the 'Recording Control' panel, and then close the panel.

Transmitting Setup

On the IZ8BLY menu, select the icon with a slider and a loudspeaker, to open the **Windows Volume Control** Mixer panel (third icon from the left in Fig

DIGITAL MODES FOR ALL OCCASIONS

Fig D.16: Setting the transmit levels

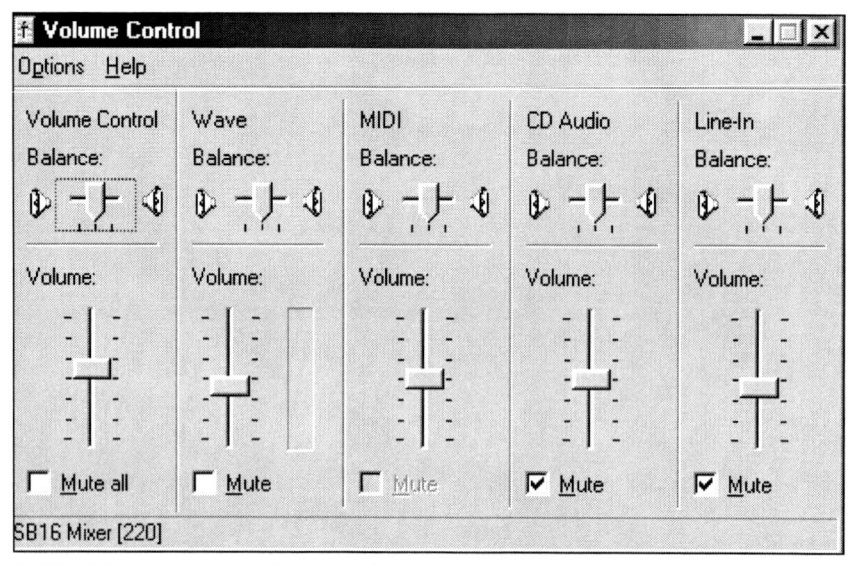

D.12). The panel name is confusing – it would be better labelled 'Playback Control'. In our case, this is the Transmit control panel. Make sure the required inputs (**Wave** and **Line In**) are available, and if necessary use **Options/Properties** to enable the correct playback control. This **Volume Control** panel controls the level of signals that come out of the sound card (**Fig D.16**).

The **Line In** signal (from the receiver) can be sent to the speaker, so you can monitor the receiver via the computer speakers. (On the right in Fig D.16). The trouble is that by sending the receiver audio to the speakers, it also sends the receiver audio into the microphone input and can trip the VOX when you don't want it to. However it is a useful way to check that the audio from the receiver is present.

Some judicious twiddling with the levels will sometimes arrive at a suitable compromise for VOX operation with **Line In** enabled, but it is safer to use only the radio speaker, and click the **Mute** check box on the **Volume Control** panel to mute the Line In output. Note that this input is muted in Fig D.16 (bottom right corner).

The sound from the software to the computer comes out the Wave output. Ensure that the Wave control is not muted, and adjust the **Wave** slider to about half way up. These are shown second control from the left in Fig D.16. The **Wave** control and **Main Volume** control (on the left in Fig D.16) both adjust the output to the rig. Set the Main Volume control (the one on the left) to the normal level you use for other applications. Remember that this control is the same as the one available in the System Tray.

Disable the transmitter (disconnect the cable from the sound card, or turn off the VOX). Check that the frequency display under the waterfall still reads 1000. Press the **Tune** button on the IZ8BLY screen and carefully monitor the tone at the PC speakers. Preferably listen to the **Line Output** with high impedance headphones, or best of all an oscilloscope.

Adjust the **Wave** control up and down, checking for a soft but 'dull' sounding tone. On the oscilloscope, check for a level with no clipping. If the tone

186

sounds at all 'bright', or the signal has a flat top or bottom to the sine wave, lower the **Wave** level until it sounds 'dull' or ceases to clip. If in doubt, lower it a bit further. This adjustment ensures that the output of the wave generator is not clipping. If you can, monitor the signal on an oscilloscope, or use another computer running a spectrogram program, and check for the level of second and third harmonics – about 2kHz and 3kHz. There should be very little. Clipping is usually caused by too much output level from the sound card, or too much load. If the clipping is asymmetrical, there will be significant output at 2kHz, which cannot be removed by the transmitter filters.

Check that the resistors in the transmit cable are not placing too much load on the sound card. Alternatively, use the **Spkr Output** socket from the sound card, which may have slightly more distortion, but will handle lower load resistances.

Now, with the rig on dummy load, tune up the transmitter on CW as you normally would, and note the ALC indication. Connect the transmit cable connected to the microphone input and the sound card, turn the rig's **VOX** on, or connect the PTT interface, switch to SSB and wind the microphone gain to minimum. Press **Tune** on the IZ8BLY screen, and increase the microphone gain slowly until you reach the normal full power output (same ALC indication as before). With any luck, this will occur at about the normal microphone gain setting for SSB. If a small adjustment in the Wave level or microphone level is not sufficient, change the shunt resistor (across the left side of the transformer in the upper half of Fig 6.4).

Do not reduce the series resistor value, as the risk of clipping will increase. Most rigs have sufficient sensitivity for there to be plenty of gain to spare. The optimum setting is so that the sound-card **Wave** level is about half way up when the ALC starts to act.

Set the transmit level (with the microphone gain) so that with the Tune signal the transmitter is at almost full output, but with no ALC indication. Then note down all the settings on the **Volume Control** pane and note the microphone gain setting. Check that the VOX or PTT operates whenever the **Tune** button is pressed, and drops out after the tone stops (after a second or two if VOX is used). Type on the keyboard and check that the VOX or PTT also operates normally. Typing slowly, the receiver should come back on between words. If all is as described, you are now ready to transmit.

> Note: It is important that the transmitter ALC does not act when operating digital modes, because the ALC is designed for SSB and will not operate correctly on peaks of the digital signals, and will result in distortion.

When using Hell, don't be concerned if it seems that your rig is only running about 20% of it's normal power as indicated on the current meter or power meter – the meter response is usually too slow to indicate the peak power. On any digital mode, don't be tempted to increase the output so that the current or power meter indicates as it would on SSB or CW. Trust the above setup to give the optimum transmit-signal level.

Finally, arrange for a friend to monitor your signal with their receiver, preferably using a spectrogram program to check for hum, unwanted sidebands and excessive signal bandwidth. Lowering the drive (microphone gain) will usually fix most problems, but it may also be necessary to lower the sound card level by reducing the **Wave** control or the **Main Volume** control. For first-time operators hum and instability are not uncommon, and usually result from RF getting back into the computer or transmitter audio. Good shielding of cables and good isolation are essential.

During the checks, make sure that your signal does not overload your friend's receiver or sound card. If necessary, they should attenuate the signal or even remove the antenna so that they monitor an S5 to S8 signal. Have the friend check for hum or noise between words (ie transmitter on but no tones sent). There should be no detectable hum or noise from the transmitter unless the signal is over S9.

Hint: The recording and playback (receive and transmit) input selections and level settings can be easily recorded for future reference by saving the control panels as graphics images (exactly as Fig D.7 and Fig D.10). With the necessary panel **in focus** (ie click on it), press **ALT+PrintScrn** (ie both at the same time). This saves the image to the clipboard. Then paste the picture into a word-processor document or graphics editor to print it or save it on disk. Fig D.7 and D.10 record the author's radio computer settings and were made in exactly this way.

Traps for Beginners

♦ Make sure the VOX is turned off when you start the PC, or everyone in the world will hear the Windows fanfare as the PC starts up!

♦ If you use Windows sounds to mark actions, you may also have other unwanted beeps and clicks transmitted. This happens when the PC speaker sounds are redirected via the sound card. It is best to turn these off within Windows (**Start/Settings/Control Panel/Sounds**). At the very least, make sure that the Amateur software does not beep at the end of each over – in the IZ8BLY software, select **File/Preferences** and on the PTT tab make sure that 'Beep on PPT off' is not ticked.

♦ Remember to turn the VOX off and remove or disable the PTT cable when you've finished operating. You would not wish to transmit the PC sounds the rest of the family cause while reading e-mail, browsing the Internet or playing games!

Adjustments for Other Software

It is very unlikely that the recording levels will need to be adjusted for other software, and if it is necessary, it is probably easiest to simply adjust the HF receiver audio gain.

The transmit levels tend to need minor adjustment. Some users report that different software does require different levels from others, and there are frequently small differences between bands as the transmitter gain changes. Multitone modes such as MT63 and MT-Hell generally have lower output. Unfortunately there is no way for the software, computer or sound card to 'remember' the optimum setting for each mode and band.

The simplest answer is to set the playback level on the mixer panel to a compromise position, and make fine adjustments with the transmitter microphone-level control. Since the power level needs adjusting to suit each QSO [3], this is hardly a problem.

References

[1] While this Appendix describes the use of 'sound cards', many new computers (and most laptop computers) have this functionality built into the main board of the computer, rather than a plug-in card. Since these built-in systems are generally compatible, the discussion that follows is equally valid for built-in sound systems.

[2] From http://iz8bly.sysonline.it/

[3] Amateurs interested in efficient use of limited Amateur band resources prefer to use the minimum power necessary for effective communication. This is also a requirement of the licence conditions in many countries.

E

How the pictures were made

In this chapter
- Photographs
- Line drawings
- Bitmap drawings
- Waveforms
- Digital mode screen captures
- Spectra and Spectrograms
- Special software captures

NUMEROUS AMATEURS have asked the author how to make good digital-mode images. Since the simple techniques for capturing images are apparently (and surprisingly) not widely known, it was felt that a small Appendix in this book, which has so many suitable examples, would be an appropriate place to divulge some simple secrets.

The techniques are all straightforward, and involve no really clever tricks or expensive software. In most cases the pictures can be easily replicated by anybody with a PC and an HF receiver. The author also uses a simple hand-held image scanner, and a selection of simple software tools and techniques. Anyone with an interest in reproducing images to be shared via e-mail, used for publication or for use as SSTV images can do the same. As an example, there is no better way to verify a DX contact using Hellschreiber (**Fig E.1**) or SSTV modes than to send a QSL card with an image of the received signal on the back!

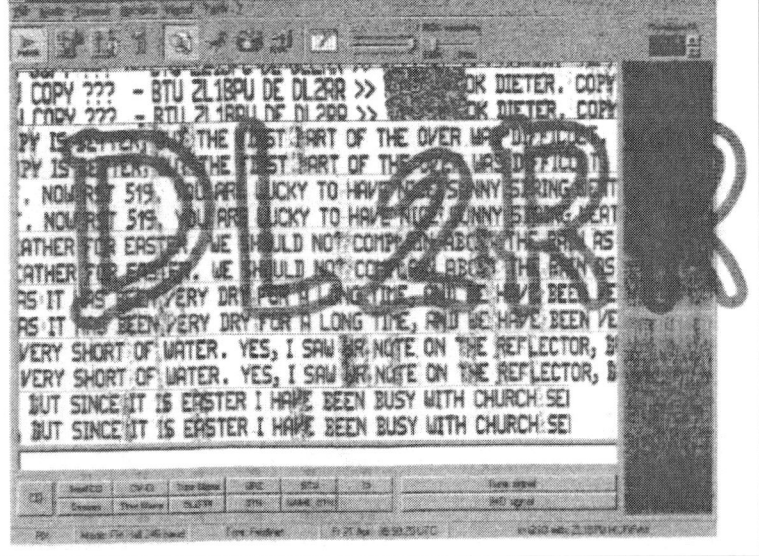

Fig E.1: A Hell contact QSL card

191

DIGITAL MODES FOR ALL OCCASIONS

Photographs

Photographs such as **Fig E.2** were scanned directly from photograph using a small hand-held scanner at 400dpi, 12-bit colour, into a graphics editor, Paintshop Pro V 4.12. After cropping to size, the pictures were saved as .tif format files. The scanner uses a Twain driver, so scanning is controlled directly from within Paintshop using the 'Acquire' function.

Other photographs were lifted directly from the Internet (with permission of course) or supplied to the author as image files, for example, Figs 5.1–Fig 5.5. To save an Internet image from Internet Explorer, **right click** on the image, select **Save Picture As**... and click on **Save** to store the image as a file. Most Internet images are in .gif or .jpg format. If there is a choice for the capture of low or high-resolution images, be sure to choose the high-resolution option when you plan to use the picture in a publication.

Fig E.2: Scanned photograph

Scanned Artifacts

Fig 5.8 was scanned from an original 1944 manual, printed in German Gothic script. Fig 5.9 (reproduced below as **Fig E.3**) was scanned from an original Hell receiver paper tape (the original was in blue ink), as was Fig 9.8, although I am assured that originally the text of this sample was much clearer. Apparently the paper used was rather absorbent, and the ink rather dilute, and therefore the ink bled into the paper soon after reception. The scanner and software used were as previously described.

One interesting point about Fig E.3: the Siemens A2 machine which transmitted the signal has no key for the 'Ø' symbol, which PA0AOB cunningly synthesised by pressing '0' and '/' at the same time!

Line Drawings

There are many ways to make line drawings for embedding in documents. The author frequently makes simple drawings using the drawing tools within Microsoft Word. Powerpoint has better drawing facilities, and there are others equally effective, such as Claris Works, which includes simple 3D modelling and shading. Specialized engineering-drawing packages such as Autocad or Visio can also be used when

Fig E.3: Scan from paper tape

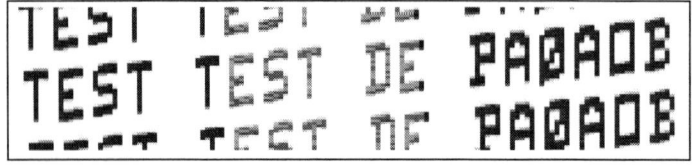

APPENDIX E: HOW THE PICTURES WERE MADE

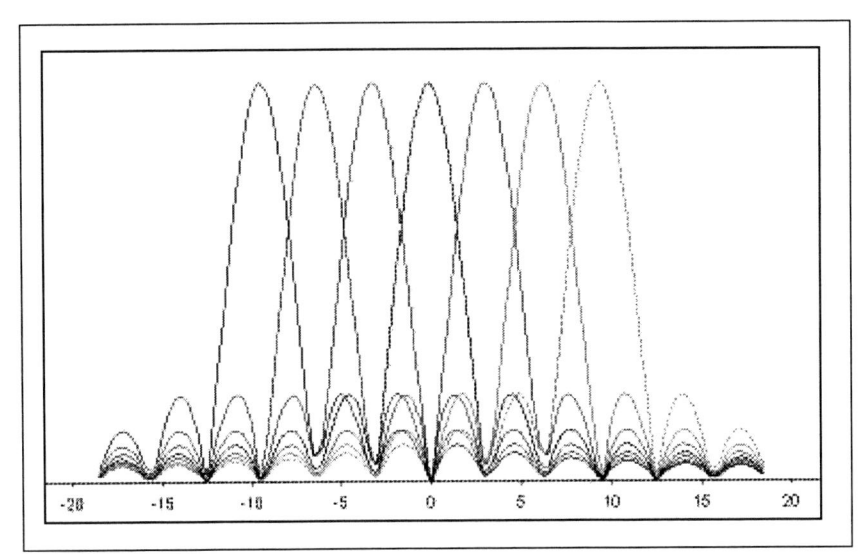

Fig E.4: A complex shape generated and graphed in Excel

more complex detail is required. The author also uses Microsoft Excel (**Fig E.4**) to mathematically describe complex shapes (such as multiple sinx/x curves depicting an MFSK signal) that can be graphed and then copied and pasted into other pictures.

Fig E.4 illustrates the spectra of several MFSK tones superimposed. A single sinx/x shape was generated in Excel, duplicated several times with different parameters, graphed and then copied and pasted to Paintshop Pro for addition of the border and transfer to a graphics file.

Many of the line-drawing tools are not capable of saving images in the desired format. Specifically, drawing packages do not as a rule save output as bit-image graphics. The pictures created in this way are line drawings (vector drawings), and are ideal where precise dimensions, clear diagrams and simple lines and shapes are required. The secret to creating .tif or other format bitmap drawings from pictures created in this way is to view the images as large as possible (ie use a high-resolution monitor and fill the screen) and then capture the image to the clipboard (use **ALT+Print Scrn**). This copies the whole of the current window (not the screen) to the clipboard.

Then paste the image into a graphics program, where it can be cut to size, rotated, edited if necessary, and saved as a bitmap image (formats .tif, .gif or .jpg). The author uses a 430mm monitor and a high-performance video card set to 1280 x 1024 x 32-bit pixel resolution. Fig E.4 and **Fig E.5** were created in this way.

It is also possible to combine the photographic and line-drawing techniques. Fig D.1 is an example. The digital photograph was first edited with Paintshop Pro to crop it, remove background, correct contrast, and to correct skew and

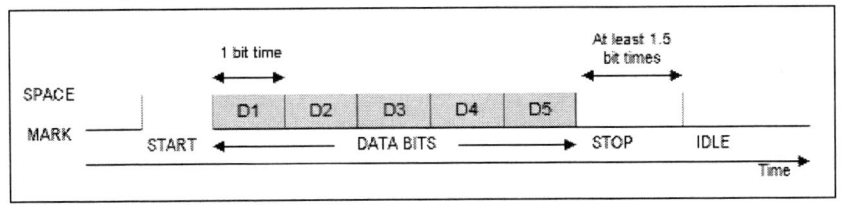

Fig E.5: A simple line drawing

193

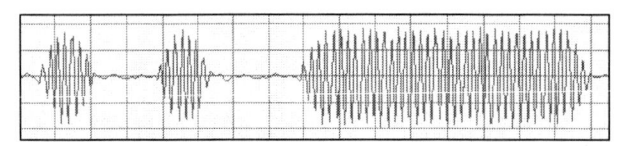

Fig E.6: Waveform from a digital audio file displayed by Goldwave

perspective effects. To ensure that no resolution was lost in the process, the image was first resampled to a much greater resolution before the corrections and clipboard copies were made. It was then imported into a Word drawing, where the text and arrows were added. Finally, the image was screen-captured to the clipboard, pasted back into Paintshop Pro, and saved as a .tif file.

Bitmap Drawings

Simple programs such as Microsoft Paint can be used to create bitmap drawings. These are pixel- or dot-based (as opposed to line- or vector-based), and so considerable skill and patience is required to generate realistic and accurate drawings. However, with the capabilities of a good bitmap-image editor such as Paintshop Pro, just about anything is possible – given enough time. The only such image in this book is Fig 3.2.

A few of the other images (photographs for example) have been 'cleaned up' by editing the scanned bitmap image at pixel level. For example, Fig 6.2 (digital photograph) had a power cord that protruded from the back of the Maplin TU on the right-hand side. This was excised, and the affected pixels were replaced with those of the adjacent background. You would need to look closely to find where this was done. Sometimes a speck of dirt or a streak from the scanner needs to be corrected in this way. Most of the images are however original and not retouched. In particular, none of the digital-mode screen shots have been retouched.

Waveforms

Wave images (rather like oscilloscope pictures) showing signal amplitude against time were created by first recording the sound as a .wav file (digital audio), using Goldwave (see **Fig E.6**). The file was then played back, displayed and viewed in zoomed mode. By zooming in far enough, it is possible to view individual samples. A very good representation of a digital oscilloscope picture can be achieved by zooming in until about 500 samples are shown (say 200–400ms), and then moving along to an interesting part of the waveform.

Once an interesting point has been found, the image is captured (**ALT+Print Scrn**) and pasted into Paintshop Pro, where it is cropped to size, the image is made negative and the colours removed. The original colours of the example in Fig E.6 were green on blue.

It is important to record the signal using a high sample rate, as high a level as possible without clipping, in order to ensure clear images. Some experimentation is necessary. It is quite difficult to record clear images in this way from live (off air) signals, because of the noise and timing problems introduced by the transmission path. It is best to use direct audio-signal samples or possibly tape recordings.

In order to record a digital file from an Amateur digital mode as audio samples, it is generally easiest to use two computers. Some computers with full duplex sound cards are capable of recording and playing (ie receiving and transmitting) at the same time, but generally it is a lot easier to set up to record from one PC to another PC, with the Line Out of the transmitting computer connect-

APPENDIX E: HOW THE PICTURES WERE MADE

ed to the Line In of the recording PC. In some cases it is possible to use a tape recorder to record the sound for later playback and analysis. Fig 10.5 was recorded in this way, and looking closely, the flutter of the tape recorder can be seen affecting the idle carrier (under the lower red line), giving it a sinusoidal characteristic. An interesting way to measure tape recorder flutter!

Digital-Mode Screen Captures

The Hellschreiber modes make the most use of screen captures for preserving interesting signals. The process is really very simple, and no different to what has already been described. With the Hell software running (such as IZ8BLY Hellschreiber), when an interesting section of text is discovered, simply press **ALT+Print Scrn** and then paste the image into a graphics editor. Crop the picture to just the area of interest, and save it as a 256 colour (8 bit) .gif or .tif file. The file size will be small if only a line or two of text is saved. Do not use .jpg format, as it causes distortion of the image edges, and is no more efficient when files are small. Fig 4.12 and Fig 9.14 (among many others) were made in this way.

Capturing image files from DOS programs is much more difficult. In general, it is best to operate in DOS and use a resident screen-shot program (the type that uses a hot-key combination to capture the screen as a file). After the screen has been captured and saved as a file, the image can be processed using a Windows-based graphics editor. A few PCs are capable of capturing graphics images from Windows DOS boxes, but in the author's experience few computers are capable of running DOS graphics applications in a DOS box, and even fewer can accurately screen-capture graphics from the DOS box. It is generally possible to screen-capture from a full-screen DOS application running under Windows, although once again this capability is very hardware dependent.

Some DOS programs can save in bitmap formats that can be read by other software. For example, PC Goes/WEFAX 3.0 can save excellent weather-satellite files in .gif format, such as shown in Fig 15.8.

SSTV pictures are quite easy to capture. Fig 1.4 and Fig 15.1 were captured from MMSSTV, using the 'Picture Viewer' pane for a large-sized view, and using **ALT+Print Scrn**, then pasting into Paintshop Pro.

Fig E.7 is another example, intended for SSTV transmission. This image has an interesting history, as the original photograph (a large 'pop-art' painting on the side of a building) was shot from below, at some distance and at a considerable angle, so considerable work was required to correct the skew, barrel distortion, varying brightness and perspective effects. The image was scanned from

Fig E.7: An image destined for SSTV

195

a photographic print at the highest-possible resolution in order to preserve image resolution throughout the corrective work.

Complete application screens are captured using **ALT+Print Scrn**, and saved as bitmaps as already described, without the need to crop or edit. Weather FAX pictures such as Fig 15.6 can also be captured in this way, since they are displayed full screen at high resolution, and can also be displayed in zoomed form.

Spectra and Spectrograms

There are many radio-signal spectra in this book. These are graphs of signal strength (vertical) and signal frequency (horizontal). In order to make comparison easy and recognition reliable, they are all recorded with the same software (which anyone can use) and with the same settings. Fig 8.2 is an example. Richard Horne's Spectrogram Version 4.2.6.4 is recommended, since it has special sampling and averaging for radio work. To replicate the images, use 5.5kHz sample rate, and an FFT size 1024, in Scope mode, with a dwell of 20ms, and a spectrum average of 128.

Once the image is tuned correctly in Spectrogram, screen-capture with **ALT+Print Scrn** and paste into a graphics editor. Make the image negative (white background) and add a rectangle to act as the graph axes. Crop the image to size, leaving only the graph, the axis graduations and the bottom and left side of the rectangle, and save it as a .gif or .tif file. The spectral images in this book also have the X-axis text 'Hz x 100' corrected to read 'Hz/100' by adding a tiny bitmap text graphic in the correct place.

The spectrogram displays frequency as one axis (vertical in Spectrogram) and time as the other. Signal strength is displayed as brightness. This type of display is especially useful for digital-mode identification. In order to make recognition of digital modes as easy as possible, the spectrograms in this book are also recorded with the same settings. Fig 8.4 is an example. Again, Richard Horne's Spectrogram Version 4.2.6.4 is recommended, using settings 5.5kHz sample rate, FFT size 1024, scroll mode (BW), dwell 20ms, and spectrum average = 1.

The user can easily replicate these images and use those in the book to help identify signals. To record the images as shown in the book, the spectrograms were screen-captured, pasted into a graphics editor, cropped to a standard size and a border added, before saving as a .tif image file.

Special Software Captures

Chirpograms such as Fig 4.15 were made using G3PLX's EVMChirp software, which has a Windows user interface with the ability to save files in .bmp bitmap form. These can be imported into a graphics editor and resaved in another format.

Dopplergrams such as Fig 4.17 and Fig 4.18 can be saved from G3PLX's EVMDop software. This software also has a .bmp file-save feature. The EVMDop settings for this technique (monitoring carrier Doppler) are typically scan width 20Hz and sample interval 5.12 seconds. Dopplergrams for recording meteorite trails (eg Fig 10.8) are recorded with wider span and shorter time intervals.

Wave editors such as Goldwave can often be used to create a synthetic sound, such as a radio signal with timing errors or Doppler shift. Fig 9.17 was created in this way. The original signal was recorded without noise as a .wav file, using

APPENDIX E: HOW THE PICTURES WERE MADE

two computers. The level was measured and the signal was then mixed with an accurate noise level, by merging files. The noise was generated mathematically in Goldwave. Finally, the image was processed during a resampling process using a mathematically-defined triangular wave to modulate the sample rate. The effect is to alter the pitch and timing of the signal slightly, replicating the effect of Doppler shift on the carrier. The image was captured as previously described (from IZ8BLY Hellschreiber), by playing the processed .wav sound back into the Hell software (within the same PC).

It is frequently possible to transmit from Amateur sound-card software and record with a Wave recorder, or to play from a Wave player into the Amateur software, but care must be taken to select the order in which the software is started in order to prevent sound-card conflicts. For example, in IZ8BLY Hellschreiber, click the 'Paper' tool to stop playback before starting the Wave recorder. Then start transmitting and record the results.

One of the most difficult things to capture is a screen shot of operating software with menu items and 'tool tips' showing. This type of thing is important for user and installation manuals. Using **Alt+Print Scrn** saves only the current window, and closes or ignores the menu pull-downs, tool tips and mouse pointer. The images in this book (Fig D.5, Fig D.6 and Fig D.8) were captured by bringing up the menu items with the mouse and pressing **Print Scrn**, which captures the complete screen to the clipboard. The image can then be pasted into an editor, cropped and saved. This technique does not allow the mouse pointer to be saved. For best results a specialised capture tool such as those used by manual writers is the best approach. This type of tool allows controlled areas to be captured, complete with mouse pointer and tool tips. Three hands would be a definite benefit for this chore – as they would for any digital mode operation!

Glossary

Analogue	As opposed to digital, analogue values smoothly vary over an infinite range of values, and can only be expressed by approximation. In computers, analogue values are expressed as binary (ie digital) approximations with a range of different resolutions.
ASK	Amplitude Shift Keying (like digital amplitude modulation).
Asynchronous	A data transmission system where small groups of data bits are transmitted so that a receiver clock can be resynchronised at the start of each group, and used to recover the data.
Antipodal	Related to the antipodes, the exact opposite side of the earth.
ARQ	Automatic Request Repeat (detected errors corrected by asking for repeats).
AX25	A packet-level data transmission protocol suited to radio use. AX25 is an adaptation of the commercial IEEE X25 protocol to allow callsigns and data repeating to be used.
Baseband	Relating to data that is not modulated on a carrier or subcarrier.
Baud	Digital-mode signalling speed, symbols/second.
Bit	A single item of binary data, 0 or 1.
bps	Bits per second. Measure of data rate.
Byte	Eight bits of binary information. Typically a number expressed in the range 0–255 which can be an alphabet character or any of a myriad of other representations. Computers and microprocessors handle bytes very effectively. Almost all wire communication and much radio communication is handled in bytes.

Carrier	A radio wave used to transport information to the receiver. The carrier is modulated in some fashion by the data.
CRC	Cyclic Redundancy Check. A complex polynomial calculation is used to derive a unique number from all the data in a message block, and this number is transmitted with the message block. When the calculation is repeated at the receiver, the data in the whole block can be assumed to be completely correct if the calculation matches the transmitted CRC. This technique is a much stronger and more reliable way of detecting errors than the use of parity or a check sum. CRC-16 (16 bits of CRC data) is the most common – CRC-32 is also used, and is extremely robust.
CSMA	Carrier Sense Multiple Access, a technique used by packet-radio modes to allow multiple users to share the same frequency.
CW	Continuous Wave. Normally implies interrupted continuous wave, as in a Morse transmission.
Digital	Relating to mathematical, logical, control or communications operations that can have two states, ie on/off, yes/no, 0/1 etc.
DPSK	Differential PSK, meaning that data is signalled by changes of phase, rather than actual phase. Almost all use of PSK modulation is differential.
DSP	Digital Signal Processing. Mathematical simulation of electrical systems such as filters, modulators and detectors, used to process radio or audio signals electronically. DSP designs cannot drift, have no temperature effects, and can have performance impossible to achieve any other way. Some designs have no hardware equivalent.
DX	Distance (typically long distance).
ECC	Error Correction Code, or any error-coding technique.
EME	Earth–Moon–Earth operating. Signals are bounced off the moon.
FAX	Facsimile, image transmission by radio.
FEC	Forward Error Correction (information used to correct text at the receiver is sent with the text).
FFT	Fast Fourier Transform. A mathematical technique used to transform data from the frequency domain to the time domain.
FSK	Frequency Shift Keying (like digital frequency modulation).
IOC	Index of Cooperation. Aspect ratio of FAX picture (width to height).
LPM	Lines per Minute. Measure of FAX transmission speed.
MCU	Multi-mode Control Unit. A radio modem which includes a microprocessor for radio protocol management and code conversion.

APPENDIX F: GLOSSARY

MFSK	Multiple Frequency Shift Keying. FSK with more than two tones.
Modem	Modulator–Demodulator. A device which converts digital signals into modulated tones for transmission by land-line or radio, and *vice versa*.
Modulation	A mathematical, electronic or electrical technique which multiplies one radio or audio signal with another. Often called a mixer or detector.
MS	Meteor Scatter.
NVIS	Near Vertical Incidence Signal (HF conditions typical on 80 and 40m). NVIS conditions are characterised by strong interaction between ground-wave and sky-wave signals with large timing differences and selective fading.
PC	Politically Correct... no, no – it also means Personal Computer, these days almost always an 'IBM PC' style computer.
PSK	Phase Shift Keying (like digital phase modulation)
QPSK	Quadrature (four phase) PSK modulation
QRL	Are you busy (is this frequency in use)? Or, this frequency/station is busy (this frequency is in use).
QRM	Interference.
QRN	Noise (typically impulse noise, static).
QRP	Low Power operation (typically 5W or less).
QRX	Please call later (or I will call you later), at a specified time (or just 'stand by!').
QRZ	Who is calling me? (ie repeat your call sign).
QSO	Conversation by radio
QSY	Can you call on ... frequency? I will call on ... frequency (or simply please change to a specified frequency).
QTH	What is your location? Or, my location is ...
RIT	Receiver Incremental Tuning. Allows the receiver frequency to be offset from the transmitter. Not a good idea for digital modes.
SSB	Single Sideband. An SSB transmitter or receiver is characterised by its ability to linearly translate all the audio components of the modulating signal to a radio frequency or *vice versa*.
SSTV	Slow Scan Television.
Subcarrier	A modulated audio carrier, a technique where an audio carrier is modulated, then translated to radio frequency by an SSB transmitter, rather than direct RF carrier modulation.
Symbol	The smallest digital mode radio signalling entity.

Synchronous	A data transmission system where data bits are transmitted at a constant rate and recovered using a clock derived from the data stream.
TCP/IP	Terminal Control Protocol/Internet Protocol. A set of communications protocols used by networks, Amateur packet networks and the Internet.
TNC	Terminal Mode Controller. A device used for packet radio which contains a device to assemble and disassemble packets, and operate the radio protocol (usually AX25). Usually also contains a modem.
TU	Terminal Unit. A modem intended for RTTY use.
Varicode	A technique where an alphabet is represented by numbers with varying numbers of bits, typically fewer bits for more frequently used letters. Normal alphabet representations have a constant bit-size.
Vocoder	Voice encoder. In digital voice transmissions, it is more efficient to code voice sounds from a catalogue of standard noises than to transmit the sound itself. The vocoder performs this job. It also usually performs the reverse process at the receiver. Better vocoders preserve the timbre, pitch and stresses of the original voice.
WPM	Words per Minute, usual method of measuring typing speed.

Bibliography

ACEC staff. 'Les téléimprimeurs, téléchiffreurs et transcodeurs ACEC – Système Coquelet', *ACEC Revue* No 3-4, 1970.

Anderson, P. *Introduction to and the Operation of AmTOR*, Kantronics Inc, 1983.

ARRL. *Specialized Communications Techniques for the Radio Amateur*, The American Radio Relay League, Inc, 1st Edition, 1975.

ARRL. *Handbook for Radio Amateurs*, The American Radio Relay League, Inc, 69th Edition, 1992.

Benson, D. & Teller, S. 'A panoramic transceiving system for PSK31', *QST*, June 2000.

Bickley, H.D. *More About Hellschreiber*, privately published, date unknown.

Bold, G.E.J. 'New sound-card aided communication modes', *ENZCON Conference Proceedings*, 2000.

Brain, C. 'Practical HF vigital voice', *QEX* May/June 2000, also available at http://www.arrl.org/qex/brain.pdf

CCIR. 'Direct printing telegraph equipment in the Maritime Mobile Service', *CCIR 476-1 1970–1974, Annex I, ITU-R*.

Cook, S. 'Hellschreiber, what it is and how it works', *Radio Communication*, April 1981.

Deutschen Zentraldruckerei. 'Der Feldfernschreiber', (technical manual for Siemens model A1/A2 Feldschreiber), *Deutschen Zentraldruckerei, Berlin D 748/1* April 1941.

Devaux, L & Smets, F. 'A seven-frequency radio-printer', *L. Les Laboratoires, Le Matériel Téléphonique*, Paris, *Electrical Communication*, 1937.

Ege G. & Promnitz H. 'Der Siemens-Hell-Feldschreiber', *Hell Techniche Mitteilungen no. 1/1940* pp. 11–20.

Ford, S. 'PSK31 – has RTTY's replacement arrived?', *QST* May 1999.

Ford, S. 'PSK31 2000', *QST* May 2000.

Ford, S. *HF Digital Handbook*, ARRL, 2001. ISBN 0-87259-823-3.

Gibbs, A. 'PSK31 the easy way', *WIA Amateur Radio*, March-June 2000.

Hal Communications. 'Clover II waveform and protocol', *Hal Communications Corp E2006 Rev A*, 1997.

Hell, R. 'Die Entwicklung des Hell-Schreibers', *Hell Techniche Mitteilungen no. 1/1940* pp. 2–11.

Hoot, J.E. *PC Goes/Wefax 3.0 User's Reference*, Software Systems Consulting, 1990.

König, R. 'The Hellschreiber', *Radio Bygones* No. 51, Feb/Mar 1998.

Martinez, J.P. 'AmTOR, an improved radioteletype system, using a microprocessor', *Radio Communication*, August 1979.

Martinez, J.P 'AmTOR, the easy way', *Radio Communication*, June/July 1980.

Martinez, J.P *AmTOR Mode B*, BARTG publication, about 1980.

Martinez, J.P 'PSK31: A new radio-teletype mode with a traditional philosophy', available at http://det.bi.ehu.es/~jtpjatae/pdf/p31g3plx.pdf.

Martinez, J.P 'The Chirps project: a new way to study HF propagation', *Radio Communication*, July– August 2000.

McDermott, T. *Wireless Digital Communications: Design and Theory*, Tucson Amateur Packet Radio Corporation (TAPR), 1996. ISBN 0-9644707-2-1.

Murray, D. *Murray Printing Telegraph*, M.A. Sydney (Ed.), Unwin, 1905.

Radiocommunications Agency. '*BR68 – Amateur Radio Licence Terms, Provisions and Limitations*', Radiocommunications Agency Booklet BR68. See http://www.radio/gov.uk/publication/ra_infor/br68/br68.htm

Ralphs, J.D. et al. 'Multi-tone signalling system employing quenched resonators for use on noisy radio-teleprinter circuits', *Proc. IEE* Vol 110 No. 9, September 1963.

Shannon, C. 'A mathematical theory of communications.' *Bell System Technical Journal*, July 1948.

Shannon, C. 'Communication in the Presence of Noise', *Proc IRE*, 1949.

Siemens & Halske AG. 'Der Siemens-Hell-Schreiber', in *Siemens Fernmelde Technik SH 8354.443.TT1*, 1943.

Stolz, A. *The Soundblaster Book*, Abacus, ISBN 1-55755-181-2.

US Office of Spectrum Management. 'Guidance for determination of necessary bandwidth', Annex J, 9/2000. See www.ntia.doc.gov/osmhome

US War Department. 'Tape facsimile equipment RC-58-B', *War Department Technical Manual TM 11-374*. US War Department 23 February 1944.

WIA Group. *PacTOR, a Short System Description*, The WIA Group, April 1991.

Zimmerman, R. 'Stand der Siemens-Hell-Fernschreibtechnik', in *Siemens Fernmelde Technik SH 7997.0,5..8.40.TT1*, Siemens & Halske AG, 1940. Reprinted in Techniche Mitteilungen der Fernmeldewerks Abteilung für Telegrafengerät, May 1940.

Index

*Chapter and appendix headings are shown in **bold***

A
Advanced Digital Modes 137
Alphabet and Code Reference .161
Amateurs involved 50
AmTOR 71
AmTOR mode A 73
AmTOR mode B 74
AmTOR, operating 75
AmTOR, performance 75
Analysis tools, other 146
ASK modulator 149

B
Bibliography 203

C
Clover .77
Clover 2000 80
Clover II modulation 78
Clover, operating 80
Complex modulation 155
Connecting up 57

D
Data synchronism 25, 156
Data codes 22
Data, considering the 19
Designs, amateur 51
Digital Fundamentals 15
Digital mode, what is? 7
Digital Modes & the Ionosphere .31
Digital modes, advantages of 9
Digital modes, disadvantages of . .10
Digital Modulation Techniques .149
Digital receiver, the 18
Digital transmitter, the 15
Doppler modulation effects 40
Dopplergrams 144

E
Encryption techniques
Error coding systems 52
Error correction techniques 26
Error correction, AmTOR
Error correction, Clover 79

F
Facsimile 133
Feld-Hell 84
FSK441 100
FSK modulator 152
Fuzzy modes 8

207

G
Getting Started 53
Glossary 199

H
Hell, how it works 83
Hell, multi-tone 85
Hell, other modes 89
Hell, PSK 87
Hellschreiber 81
HF performance, designing modes for .42
History .45
How the Pictures were Made . .191

I
Image Modes 129
Image Modes, new 136
Image transmission 49
Introduction 5
Introduction to the modes 11
Ionospheric sounding 140
ITA2 161
ITA5 . . . . .163, 164

M
MFSK Modes 91
MFSK, how it works 93
MFSK16 94
MFSK8 98
Moore code 162
Morse .165
MT63 .103
MT63 software 108
MT63, receiving 106
MT63, transmitting 105
Multimode programs 171
Multipath reception 36

N
Noise and interference 31

O
Operating considerations, special .67
Operating hints
Oscilloscopes 144
Other Tools for Amateurs 143

P
PacTOR 109
PacTOR ARQ
PacTOR FEC 102

PacTOR II 102
PC-ALE 137
PC Sound Card 175
Performance, Clover 79
Performance, comparative 42
PSK modulator 154
PSK, other modes 119
PSK31 .115
PSK31, how it works 116
PSK31, operating 117
PSK31, software for 118
PSK31 varicode 166

Q
Q15X25 139

R
Reception, automatic 46
Resources 172
RTTY .121
RTTY described 123
RTTY history 121
RTTY, hardware for 127
RTTY, software for 128
RTTY, using 125

S
Satellite images 134
Serial Data Transmission 25
Setting up 61
Software & Information Sources . .169
Software, choosing 66
Software, digital modes 170
Spectrograms & spectrum analysers ..143
SSTV .129
SSTV modes and VIS codes 167
STANAG 4285, 4529 139
Sound card, inside the 175
Sound card setup for digital
  modes 180

T
Text transmission 46
Throb 99

V
Voice communications digital . .140

W
Wave tools 145
What you need 53